W9-AEL-920

surfing through hyperspace

Previous Works by Clifford A. Pickover

The Alien IQ Test

Black Holes: A Traveler's Guide

Chaos and Fractals: A Computer-Graphical Journey

Chaos in Wonderland: Visual Adventures in a Fractal World

Computers and the Imagination

Computers, Pattern, Chaos, and Beauty

Fractal Horizons: The Future Use of Fractals

Frontiers of Scientific Visualization (with Stu Tewksbury)

Future Health: Computers and Medicine in the 21st Century

The Girl Who Gave Birth to Rabbits

Keys to Infinity

The Loom of God

Mazes for the Mind: Computers and the Unexpected

Mit den Augen des Computers

The Pattern Book: Fractals, Art, and Nature

The Science of Aliens

Spider Legs (with Piers Anthony)

Spiral Symmetry (with Istvan Hargittai)

Strange Brains and Genius

Time: A Traveler's Guide

Visions of the Future: Art, Technology, and Computing in the 21st Century

Visualizing Biological Information

surfing through hyperspace

Understanding Higher Universes in Six Easy Lessons

Clifford A. Pickover

New York Oxford

OXFORD UNIVERSITY PRESS

1999

Oxford University Press

Oxford   New York

Athens   Auckland   Bangkok   Bogotá   Buenos Aires   Calcutta
Cape Town   Chennai   Dar es Salaam   Delhi   Florence   Hong Kong   Istanbul   Karachi
Kuala Lumpur   Madrid   Melbourne   Mexico City   Mumbai
Nairobi   Paris   São Paulo   Singapore   Taipei   Tokyo   Toronto   Warsaw
and associated companies in
Berlin   Ibadan

Copyright © 1999 by Clifford A. Pickover

Published by Oxford University Press, Inc.
198 Madison Avenue, New York, NY 10016

Oxford is a registered trademark of Oxford University Press

All rights reserved. No part of this publication may be reproduced,
stored in a retrieval system, or transmitted, in any form or by any means,
electronic, mechanical, photocopying, recording, or otherwise,
without the prior permission of Oxford University Press.

Library of Congress Cataloging-in-Publication Data
Pickover, Clifford A.
Surfing through hyperspace : understanding higher universes in six
easy lessons / by Clifford A. Pickover.
p.   cm.
Includes bibliographical references and index.
ISBN 0–19–513006–5
1. Cosmology.
2. Hyperspace.
3. Fourth dimension.
4. Science—Philosophy
5. Mathematics—Philosophy.
I. Title.
QB981.P625   1999
523.1—dc21   98–48660

1 3 5 7 9 8 6 4 2

Printed in the United States of America
on acid-free paper

29.

This book is dedicated to
Gillian Anderson, David Duchovny, and Chris Carter

acknowledgments and disclaimers

Dost thou reckon thyself only a puny form
When within thee the universe is folded?

—Baha'u'llah quoting Imam Ali, the first Shia Imam

I owe a special debt of gratitude to mathematician Dr. Rudy Rucker for his wonderful books and papers from which I have drawn many facts regarding the fourth dimension. I heartily recommend his book *The Fourth Dimension* for further information on higher dimensions in science and spirituality. I also thank Dr. Thomas Banchoff, author of *Beyond the Third Dimension*, for his pioneering work in visualizing the fourth dimension. Among his many talents, Dr. Banchoff is also an expert on the nineteenth-century classic *Flatland*, which continues to be an excellent introduction to the interrelationship between worlds of different dimensions. The various works of Martin Gardner, listed in the Further Readings section, have also been influential in my formulating an eclectic view of the fourth dimension. I thank Kirk Jensen, my editor at Oxford University Press, for his continued support and encouragement, and Brian Mansfield, Lorraine Miro, Carl Speare, Arlin Anderson, Clay Fried, Gary Adamson, Ben Brown, Sean Henry, Michelle Sullivan, Greg Weiss, and Dan Platt for useful advice and comments. Brian Mansfield prepared many of the illustrations and April Pedersen drew the wonderful small cartoons used in the chapter openings on pages 8, 43, 69, 96, 129, 155, and 163. Some of the drawings of Earthly animals, such as the seashells and trilobites, come from the Dover Pictorial Archive; one excellent source is Ernst Haeckel's *Art Forms in Nature*.

Many of the science-fiction books listed in Appendix B were suggested by Dr. Sten Odenwald. The twisted bottle in Figure 5.10 is courtesy of artist/writer Paul Ryan of the Earth Environmental Group and was drawn by Gary Allen. Figures 3.3, 3.8b, and 5.2 and are courtesy of the National Library of Medicine's Visible Human Project. Don Webb is the author of the poem, "Reflections on a Tesseract Rose."

The Chinese calligraphy in the Introduction was contributed by Dr. Siu-Leung Lee, who has been practicing the art of calligraphy for more than forty years. Capable of writing in many styles, Dr. Lee has created his own form evolving from those of the Han and Jin dynasties. Roughly translated, his calligraphy is: "We surf in higher dimensions." The calligraphy uses lettering that combines archaic structure and fluid movements to symbolize the dynamic nature of the universe.

This book was not prepared, approved, or endorsed by any entity associated with the Federal Bureau of Investigation, nor was it prepared, approved, licensed, or endorsed by any entity involved in creating or producing the *X-Files* TV show.

An unspeakable horror seized me. There was a darkness; then a dizzy, sickening sensation of sight that was not like seeing; I saw a Line that was no Line; Space that was not Space; I was myself, and not myself. When I could find voice, I shrieked aloud in agony, "Either this is madness or it is Hell." "It is neither," calmly replied the voice of the Sphere, "it is Knowledge; it is Three Dimensions; Open your eye once again and try to look steadily."

—Edwin Abbott Abbott, *Flatland*

Even the mathematician would like to nibble the forbidden fruit, to glimpse what it would be like if he could slip for a moment into a fourth dimension.

—Edward Kasner and James Newman,
Mathematics and the Imagination

May I pass along my congratulations for your great interdimensional breakthrough. I am sure, in the miserable annals of the Earth, you will be duly inscribed.

—Lord John Whorfin
in *The Adventures of Buckaroo Banzai Across the 8th Dimension*

A man who devoted his life to it could perhaps succeed in picturing himself a fourth dimension.

—Henri Poincaré, *"L'Espace et la géométrie"*

contents

preface

To consider that after the death of the body the spirit perishes is like imagining that a bird in a cage will be destroyed if the cage is broken, though the bird has nothing to fear from the destruction of the cage. Our body is like the cage, and the spirit like the bird. We see that without the cage this bird flies in the world of sleep; therefore, if the cage becomes broken, the bird will continue and exist. Its feelings will be even more powerful, its perceptions greater, and its happiness increased.

—Abdu'l-Baha, *Some Answered Questions*

The bird fights its way out of the egg. The egg is the world. Who would be born must first destroy a world. The bird flies to God. That God's name is Abraxas.

—Hermann Hesse, *Demian*

Touring Higher Worlds

I know of no subject in mathematics that has intrigued both children and adults as much as the idea of a fourth dimension—a spatial direction different from all the directions of our normal three-dimensional space. Philosophers and parapsychologists have meditated on this dimension that no one can point to but may be all around us. Theologians have speculated that the afterlife, heaven, hell, angels, and our souls could reside in a fourth dimension—that God and Satan could literally be lumps of hypermatter in a four-dimensional space inches away from our ordinary three-dimensional world. Throughout time, various mystics and prophets have likened our world to a three-dimensional cage[1] and speculated on how great our perceptions would be if we could break from the confines of our world into higher dimensions. Yet, despite all the philosophical and spiritual implications of the fourth dimension, this extra dimension also has a very practical side. Mathematicians and physicists use the fourth dimension every day in calculations. It's part of important theories that describe the very fabric of our universe.

I first became excited about the possibility of a fourth dimension as a child watching a TV rerun of the 1959 science-fiction movie *The 4D Man*. This clever thriller described the adventures of a scientist who develops a method of transposing matter, enabling him to pass through walls, windows, water, and women. Here is a snippet of the movie's dialogue:

> *Scott Nelson:* That's what you've done with your force field. You've compressed the energy of years into a moment.
> *Linda Davis:* But . . . that's like . . . the fourth dimension.
> *Captain Rogers:* I don't believe it. I'm a cop. I work with facts. Now I have to start looking for something that saps the life out of a man like juice out of an orange.
> *Tony Nelson:* Nothing can stop him. Can't imprison him or surround him with men or guns or tanks. No walls thick enough or guns strong enough. A man in the fourth dimension is indestructible.

The movie has a bevy of Hollywood stars—Patty Duke and Lee Meriwether, just to mention two. The plot involved a scientist discovering a dimension in which he can walk through solid matter. I hope I'm not ruining the movie by telling you the bizarre ending where he materializes out of the fourth dimension into our three-dimensional world while passing through a brick wall. Ouch!

You can't imagine how profoundly affected I was by the blurring of fact and fiction. To a young boy, the strange array of physical and mathematical ideas made the unbelievable seem a frighteningly real possibility. I knew that if an accessible fourth dimension existed, it would actually be possible to escape from a prison by temporarily going into the fourth dimension—like a bird leaving its nest for the first time, flying upward, and joyfully revelling in its newly found third dimension.

My fascination with the fourth dimension was later stimulated by Steven Spielberg's 1982 movie *Poltergeist* in which a family living in a suburban development is faced with menacing phenomena: a child who disappears, furniture that moves by itself, and weird powers gusting through the house and frightening everyone.

Do any of you recall the *Poltergeist* scene in which balls are thrown into a closet and then seem to magically reappear from the ceiling in another location in the house? This could easily be explained if the ball took a route through the fourth dimension—as you will learn later in this book. Even the early 1960s TV show *The Outer Limits* touched on higher dimensions. In one particularly poignant episode, a creature from the Andromeda galaxy lived in a higher dimension than ours and was pulled into our universe as a result of terrestrial experiments with a new form of three-dimensional TV. Although the creature is both wise and friendly, its visit to our world causes quite a pandemonium.

For decades, there have been many popular science books and science-fiction novels on the subject of the fourth dimension. My favorite science book on the subject is Rudy Rucker's *The Fourth Dimension*, which covers an array of topics on space and time. My favorite science-fiction story is Robert Heinlein's "—And He Built a Crooked House," first published in 1940. It tells the tale of a California architect who constructs a four-dimensional house. He explains that a four-dimensional house would have certain advantages:

> I'm thinking about a fourth spatial dimension, like length, breadth, and thickness. For economy of materials and convenience of arrangement you couldn't beat it. To say nothing of ground space—you could put an eight-room house on the land now occupied by a one-room house.

Unfortunately, once the builder takes the new owners on a tour of the house, they can't find their way out. Windows and doors that normally face the outside now face inside. Needless to say, some very strange things happen to the terrified people trapped in the house.

Many excellent books on the fourth dimension, are listed in the Further Readings at the end of this book. So, why another book on higher-dimensional worlds? I have found that many previous books on this subject lacked an important element. They don't focus wholeheartedly on the physical appearance of four-dimensional beings, what mischief and good they could do in our world, and the religious implications of their penetration into our world. More important, many prior books are also totally descriptive with no formulas for readers to experiment with—not even simple formulas—or are so full of complicated looking equations that students, computer hobbyists, and general audiences are totally overwhelmed.

The fourth dimension need not remain confined to Hollywood and the realm of science fiction, beyond the range of exciting experiment and careful thought. Many of the ideas, thought exercises, and numerical experiments in this book are accessible to both students and seasoned scientists. A few pieces of computational recipes are included so that computer hobbyists can explore higher-dimensional worlds. But those of you with no interest in computing can easily skip these sections and investigate the mental realms, unaided by computation. In this book, I'll discuss such concepts as "degrees of freedom" and then gradually work my way up to more sophisticated concepts such as the possibility of stuffing huge whales into tiny four-dimensional spheres. The Appendices discuss a number of stimulating problems—from a four-dimensional version of Rubik's cube and the evolution of four-dimensional biologies, to four-dimensional fractal quaternions with infinitely complex structures. However, the emphasis will be on the powers and appearances of four-dimensional beings. I want to know if humankind's Gods could exist in the fourth dimension.[2]

If a fourth dimension did exist, God could be so close we could hear His breath, only inches away, but impossible to see because our perception is seemingly confined to three dimensions.

What if *you* could visit a four-dimensional world filled with intelligent life-forms? Would the aliens have heads, arms, and legs, or even be vaguely humanoid? What capabilities would they possess if they were to visit our world? The challenging task of imagining beings from other dimensions is useful for any species that dreams of understanding its place in a vast universe with infinite possibilities.

Is there really a fourth dimension we can explore and understand? This question is an old one posed by philosophers and scientists and has profound implications for our worldview. There seems to be no reason why a four-dimensional world of material four-dimensional objects could not exist. The simple mathematical methods in this book reveal properties of shapes in these higher spaces, and, with special training, our dimensionally impoverished minds may be able to grasp the "look and feel" of these shapes. As we speculate, we touch on the realm of mysticism and religion—because in the fourth dimension the line between science and mysticism grows thin.

Hyperbeings living in a four-dimensional space can demonstrate the kinds of phenomena that occur in hyperspace. For example, a hyperbeing can effortlessly remove things before our very eyes, giving us the impression that the objects simply disappeared. This is analogous to a three-dimensional creature's ability to remove a piece of dirt inside a circle drawn on a page without cutting the circle. We simply lift the dirt into the third dimension. To two-dimensional beings confined to the piece of paper, this action would appear miraculous as the dirt disappeared in front of their eyes. The hyperbeing can also see inside any three-dimensional object or life-form, and, if necessary, remove anything from inside. As we will see later, the being can look into our intestines, examine our nervous system, or remove a tumor from our brain without ever cutting through the skin. A pair of gloves can be easily transformed into two left or two right gloves, and three-dimensional knots fall apart in the hands of a hyperbeing who can lift a piece of the knot up into the fourth dimension. In his correspondence and verse, the eminent physicist James Clerk Maxwell referred to the fourth dimension as the place where knots could be untied:

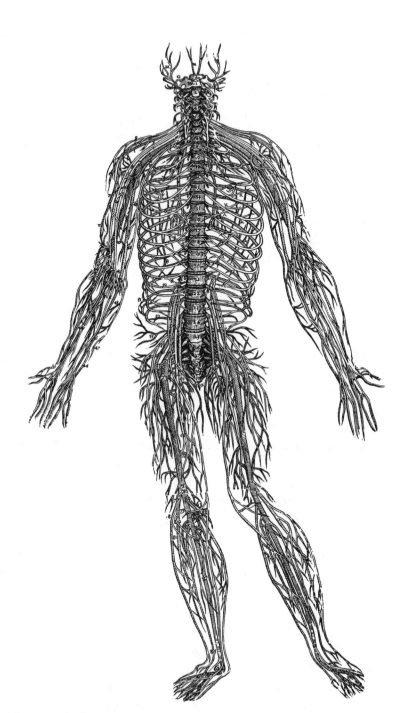

A being in the fourth dimension could see all our nerves at once or look inside our intestines and remove tumors without ever cutting our skin. The powers of these "hyperbeings" are a central topic of this book.

My soul is an entangled knot
Upon a liquid vortex wrought
The secret of its untying
In four-dimensional space is lying.

I call such four-dimensional beings "Gods." If we ever encounter beings that can move in a fourth spatial dimension, we would find that they can perform levitation, bloodless surgery, disappear in front of our eyes, walk through walls. . . . It would be very difficult to hide from them no matter where we went. Objects locked in safes would be easy for them to retrieve. If such a being were observed in biblical times, it would be considered a God with many characteristics of omniscience, omnipresence, and omnipotence.

A Note on Terminology

Most Earthly cultures have a vocabulary with words like up, down, right, left, north, south, and so forth. Although the terms "up" and "down" have meaning for us in our three-dimensional universe, they are less useful when talking about movements from the three-dimensional universe into the fourth dimension. To facilitate our discussions, I use the words "upsilon" and "delta," denoted by the Greek letters Υ and Δ. These words can be used more or less like the words up and down, as you will see when first introduced to the terms in Chapter 3.

The term "hyperspace" is popularly used when referring to higher dimensions, and *hyper-* is the correct scientific prefix for higher-dimensional geometries. Like other authors, I have adhered to the custom of using hyperspace when referring to higher dimensions. The word "hyperspace" was coined by John W. Campbell in his short story "The Mightiest Machine" (1934), and the term has been used both by science-fiction writers and physicists ever since. Moreover, physicists sometimes use hyperspace in discussions of the structure of our universe. For example, if we cannot move faster than light in this universe, perhaps we can take a shortcut. Astrophysicists sometimes speculate that there may be a way to slip out of ordinary space and return to our own universe at some other location via a crumpling of space. This severe folding takes place in hyperspace so that two seemingly faraway points are brought closer together. Some physicists also view hyperspace as a higher dimension in which our entire universe may be curved—in the same way that a flat piece of paper can be flexed or rolled so that it curves in the third dimension.

Not only may hyperspace play a role on the galactic and universal size scale, but it may help characterize the ultrasmall. Physicist John A. Wheeler has suggested that empty space may be filled with countless tiny wormholes connecting different parts of space, like little tubes that run outside of space and back in again at some distant point. Wheeler describes these wormholes as running through "superspace," which seems similar to what science fiction has called hyperspace for over a half-century.

The FBI's Four-Dimensional Smorgasbord

This book will allow you to travel through dimensions—and you needn't be an expert in physics, mathematics, or theology. Some information is repeated so that each chapter contains sufficient background data, but I suggest that you read the chapters in order as you gradually build your knowledge. I start most chapters with a dialogue between two quirky FBI agents who experiment with the fourth dimension from within the (usually) safe confines of their FBI office in Washington, D.C. You are Chief Investigator of four-dimensional phenomena. Your able student is a novice FBI agent initially assigned to work with you to debunk your outlandish theories. But she gradually begins to doubt her own skepticism. This simple science fiction is not only good fun, but it also serves a serious purpose— that of expanding your imagination. We might not yet be able to easily travel into the fourth dimension like the characters in the story, but at least the fourth dimension is not forbidden by the current laws of physics. I also use science fiction to explain science because, over the last century, science fiction has done more to communicate the adventure of science than any physics book. As you read the story, think about how humans might respond to future developments in science that could lead to travel in a fourth dimension.

When writing this book, I did not set out to create a systematic and comprehensive study of the fourth dimension. Instead, I have chosen topics that interested me personally and that I think will enlighten a wide range of readers. Although the concept of the fourth dimension is more than a century old, its strange consequences are still not widely known. People often learn of them with a sense of awe, mystery, and bewilderment. Even armed with the mathematical theories in this book, you'll still have only a vague understanding of the fourth dimension, and various problems, paradoxes, and questions will plague you. What are the chances that we could learn to communicate with a four-dimensional extraterrestrial? Would they have internal organs like our own? We'll encounter all these and other questions as we open doors.

I've attempted to make *Surfing Through Hyperspace* a strange journey that unlocks the doors of your imagination with thought-provoking mysteries, puzzles, and problems on topics ranging from hyperspheres to religion. A resource for science-fiction aficionados, a playground for philosophers, an adventure and education for mathematics students, each chapter is a world of paradox and mystery.

I hope my army of illustrators will also stimulate your imagination in ways that mere words cannot. Imagery is at the heart of much of the work described in this book. To better understand and contemplate the fourth dimension, we need our eyes. To help visualize higher-dimensional geometrical structures like hypercubes, I use computer graphics. To help visualize higher-dimensional beings, I recruit artists from different backgrounds to produce visual representations from myriad perspectives. For many of you, seeing hypothetical four-dimensional beings, and their intersections with our ordinary three-dimensional world, will clarify concepts. I often use the technique of explaining phenomena in lower dimensions to help understand higher dimensions.

Why contemplate the appearance of four-dimensional beings and their powers? Mathematicians and artists feel the excitement of the creative process when they leave the bounds of the known to venture far into unexplored territory lying beyond the prison of the obvious. When we imagine the powers of hyperbeings, we are at the same time holding a mirror to ourselves, revealing our own prejudices and preconceived notions. The fourth dimension appeals to young minds, and I know of no better way to stimulate students than to muse about higher-dimensional worlds. Creative minds love roaming freely through the spiritual implications of the simple mathematics.

Could creatures be hiding out in the fourth dimension at this very moment observing us? If you had the opportunity of stepping off into the fourth dimension, even for a few minutes, and looking down on our world, would you do it? (Before answering, remember that you would be peering into the steaming guts of your best friends. You'll learn more about this unavoidable X-ray vision effect later.) None of these questions can be answered to scientists', theologians,' or psychologists' satisfaction. Yet the mere asking stretches our minds, and the continual search for answers provides useful insights along the way.

As in all my previous books, you are encouraged to pick and choose from the smorgasbord of topics. Many chapters are brief to give you just the tasty flavor of a topic. Those of you interested in pursuing specific topics can find additional information in the referenced publications. To encourage your involvement, the book is loaded with numerous what-if questions for further thought. Spread the spirit of this book by posing these questions to your students; to your priest, rabbi, mullah, or congregation; to your buddies at the bowling alley and local shopping mall; to your family the next time you plunk down on the couch to watch *The X-Files*, or when you can't seem to find your keys and wonder if they have escaped your notice by temporarily retreating into the fourth dimension.

Whatever you believe about the possibility of a fourth dimension, the dimensional analogies in this book raise questions about the way you see the world and will therefore shape the way you think about the universe. For example, you will become more conscious about what it means to visualize an abstract object in your mind.

By the time you've finished this book, you will be able to

- understand arcane concepts such as "degrees of freedom," "hyperspheres," and "tesseracts."
- impress your friends with such terms as: "enantiomorphic," "extrinsic geometry," "quaternions," "nonorientable surfaces," "Kaluza-Klein theory," and "Hinton cubes."
- write better science-fiction stories for shows such as *Star Trek*, *The X-Files*, or *The Outer Limits*.
- conduct computer experiments dealing with several aspects of the fourth dimension.
- understand humanity's rather limited view of hyperspace, and how omniscient gods could reside in the fourth dimension while we are only dimly aware of their existence.
- stuff a whale into a ten-dimensional sphere the size of a marble.

You might even want to go out and buy a CD of the music from the *X-Files* TV show.

Let me remind you—as I do in many of my books—that humans are a moment in astronomic time, a transient guest of the Earth. Our minds have not sufficiently evolved to comprehend all the mysteries of higher dimensions. Our brains, which evolved to make us run from predators on the African grasslands, may not permit us to understand four-dimensional beings or their thought processes. Given this potential limitation, we hope and search for knowledge and understanding. Any insights we gain as we investigate structures in higher dimensions will be increasingly useful to future scientists, theologians, philosophers, and artists. Contemplating the fourth dimension is as startling and rewarding as seeing the Earth from space for the first time.

January 1999   C.A.P.
Yorktown Heights, New York

The trouble with integers is that we have examined only the small ones. Maybe all the exciting stuff happens at really big numbers, ones we can't get our hand on or even begin to think about in any very definite way. So maybe all the action is really inaccessible and we're just fiddling around. Our brains have evolved to get us out of the rain, find where the berries are, and keep us from getting killed. Our brains did not evolve to help us grasp really large numbers or to look at things in a hundred thousand dimensions.

—Ronald Graham, quoted in Paul Hoffman's
"The man who loves only numbers"

Whoever feels the touch of my hand shall become as I am, and hidden things shall be revealed to him . . . I am the All, and the All came forth from me. Cleave a piece of wood and you will find me; lift up a stone and I am there.

—*The Gospel According to Thomas*

Preparing for hyperspace. It's rather unpleasantly like being drunk.

—Ford Prefect in *The Hitchhiker's Guide to the Galaxy*

introduction

An Ancient Grotto; Cherbourg, France; 4:00 P.M.

The year is 2012 and you are the chief FBI investigator of unsolved cases involving paranormal or unexplained phenomena. Your background in mathematics makes you especially interested in those cases that may be explained by studying the fourth dimension.

Today you are beside a shrine in Cherbourg, France, just a mile away from one of France's largest Chinese populations. The air is musty and damp as vague perpetual clouds float overhead in a fractal pattern of powder blue and gray. Occasionally you hear the cry of a blackbird.

With you is your scientifically trained partner, Dr. Sally Skinner. Sally, an FBI forensic pathologist, was initially assigned to work with you to debunk your far-out theories. But as your partnership progressed, even Sally had a hard time explaining some of the bizarre happenings you encountered in your investigations.

Sally pushes back a lock of cinnamon-colored hair that the soft breeze has teased out of place. Her gaze is intense. "Why did you bring me all this way?"

You tap your knuckles on a nearby tombstone embossed with ancient Chinese calligraphy. "Some unusual sightings have been reported here."

She puts her hands on her hips. "You dragged me all this way to France to investigate ghosts and aliens?"

You chuckle. Sally, originally trained as a physician, knows a lot about medicine but so little about the fourth dimension. She has not been keeping up-to-date on the latest top-secret research on the Omegamorphs, mysterious beings from higher dimensions that sometimes seem to penetrate your three-dimensional universe. Sure, she's heard the rumors, the tabloid gossip, but she has not fully accepted the philosophi-

cal and national security consequences of penetration from a higher world.

"What's that?" Sally stops dead in her tracks, her gaze transfixed on the nearby bushes.

There are rustling sounds, like wind on dry leaves. An ammonia odor permeates the cool air. Suddenly, several blobs of skinoid materialize.

Sally clenches her fists. "My God! What are they?"

The pulsating objects resemble flesh-colored balloons constantly changing size. They remind you of floating splotches in a lava light.

You smile. "I have the key to all the enigmas of the universe: God, ghosts, and all manner of paranormal."

Sally stares at the bobbing blobs. Some contain teeth, claws, and hair. For an instant the blobs become wormlike, with intricate vascular systems beneath their translucent coverings. "I don't want to hear any more about your philosophy. Let's get out of here."

You sit down on a cold gravestone and point to the strange shapes that float around in midair. "It's a single four-dimensional being."

Sally withdraws her .22-caliber pistol and takes cover behind an old oak tree. "How do you know that?" she whispers. With her free hand she

A 4-D being appears before you and Sally in Cherbourg, France. Although the being seems to consist of separate pieces, the various components are connected in the fourth dimension and constitute a single creature.

snaps a few photographs of the being using a miniature camera concealed in the jacket of her prim and proper suit. "How can you remain so calm!"

In front of you is a fleshy ball, the size of a large pumpkin. It bounces up and down off the ground to a height of about four feet and looks just like human skin: mostly smooth, fleshy, with an occasional wrinkle and vein.

You back up a hasty half-step. "Think of it this way. Consider a two-dimensional world resembling a sheet of paper, or the surface of a pond, with two-dimensional creatures confined to the world and gazing only along the surface. How would *you* appear to the inhabitants of such a world if you tried to interact with them?"

"They would only see a small slice of me?"

"Yes. They would only see cross sections of you as you intersected their universe. For example, your finger would look like a flat disc that grew in size as you pushed it through their world. Your five fingers might look like five separate circles. They would only see irregular shapes with skin boundaries as you entered their world. Similarly, a hyperbeing who lived in the fourth dimension would have a cross section in our space that looked like floating balloons made of skin."

"Some of them don't look like blobs of skin."

"Correct. Imagine how complicated your two-dimensional cross section would appear at the level of your ear or open mouth, especially if parts of your skin were translucent like a jellyfish."

The blobs and wormy shapes drift closer to Sally and she raises her gun.

"Don't worry, Sally. The being probably wants to pick you up. A four-dimensional being would be a God to us. It would see everything in our world. It could even look inside your stomach and remove your breakfast without cutting through your skin, just like you could remove a speck inside a two-dimensional creature by picking the speck up into the third dimension, perpendicular to the creature, without breaking the skin of the creature."

Sally backs away from the four-dimensional being. "You jerk. I don't want to hear any more—"

With those words, Sally Skinner disappears into the fourth dimension. All you can hear is the blowing wind, like the chanting of monks.

And then suddenly, the wind stops. There are no bird sounds. The oak leaves do not flutter. The blackbirds above you seem to never cry, never move. They float, with dark wings outstretched and motionless, as if suspended, forever frozen in space.

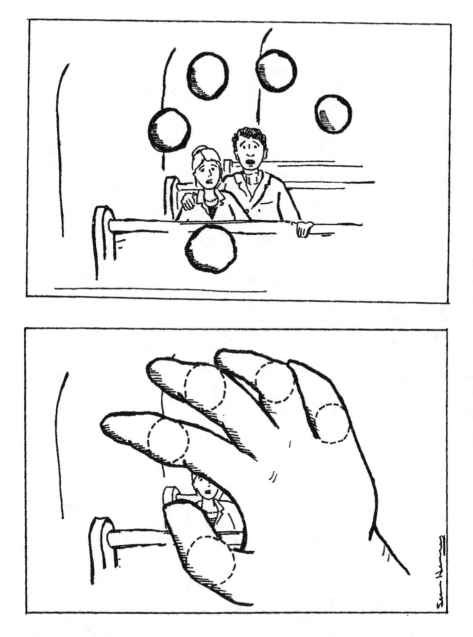

A hand from the fourth dimension could appear as five separate flesh-balls (a) to you and Sally, just as a hand intersecting a plane appears as five separate circles (b). (Drawing by Sean Henry.)

1604

Orison.

35.

surfing through hyperspace

To a frog with its simple eye, the world is a dim array of greys and blacks. Are we like frogs in our limited sensorium, apprehending just part of the universe we inhabit? Are we as a species now awakening to the reality of multidimensional worlds in which matter undergoes subtle reorganizations in some sort of hyperspace?

—Michael Murphy, *The Future of the Body*

A place is nothing: not even space, unless at its heart—a figure stands.

—Paul Dirac, *Principles of Quantum Mechanics*

Traveling through hyperspace ain't like dusting crops, boy.

—Han Solo in *Star Wars*

degrees of freedom

You have returned from Cherbourg and are relaxing in your office at the Washington Metropolitan Field Office of the Federal Bureau of Investigation located on 1900 Half Street, Washington, D.C. Very few people know your office exists because its door is cleverly disguised as an elevator bearing a perpetual "out-of-order" sign.

Inside, on the back of your door, is the colorful FBI seal and motto "Fidelity, Bravery, Integrity." The peaked beveled edge circumscribing the seal symbolizes the severe challenges confronting the FBI and the ruggedness of the organization.

Below the motto is a handmade sign that reads

I BELIEVE THE FOURTH DIMENSION IS REAL.

Sally follows you into your high-ceilinged office crammed with books and electrical equipment. Lying scattered between the sofa and chairs are three oscilloscopes, a tall Indian rubber plant, and a Rubik's cube. A small blackboard hangs on the wall. The bulletproof, floor-to-ceiling windows give the appearance of a room more spacious than it really is.

Sally eyes the electrical hardware. "What's all this?"

She nearly knocks an antique decanter from a table onto your favorite gold jacquard smoking jacket slung comfortably over a leather chair.

You don't answer her immediately but instead slip a CD into a player. Out pours Duke Ellington's "Satin Doll."

Sally snaps her fingers to get your attention.

You turn to her. "About France—"

"Yes?"

3

"I'd like to apologize for not rescuing you sooner. In less than a minute, I found you lying beside a tombstone. You weren't harmed."

She nods. "I still don't understand what happened to me."

"That's why we're here. I'm going to teach you about the fourth dimension and show you things you've never dreamed of."

She rolls her eyes. "You sound like my ex-husband. All talk."

You hold up your hand in a don't-shoot pose. "Don't worry, you'll like this. Take a seat."

You take a deep breath before starting your lecture. "The fourth dimension corresponds to a direction different from all the directions in our world."

"Isn't time the fourth dimension?"

"Time is one example of a fourth dimension, but there are others. Parallel universes may even exist besides our own in some ghostly manner, and these might be called other dimensions. But I'm interested in a fourth *spatial* dimension—one that exists in a direction different from up and down, back and front, right and left."

Sally lowers herself evenly into a chair. "That sounds impossible."

"Just listen. Our ordinary space is three-dimensional because all movements can be described in terms of three perpendicular directions." You remain standing and gaze down at Sally. "For example, let's consider the relative position of our hearts. You do have a heart, Sally?"

"Very funny."

From your desk drawer, you remove a tape measure and hand one end of it to her. "I can say that your heart is about four feet south of mine, one foot east of mine, and two feet down from mine. In fact, I can specify any location with three types of motion."

She nods, apparently growing more interested in your talk. Outside there is a crack of lightning. You look toward the sky and then back at Sally. The gathering clouds and mists reflected in her eyes make them seem like gray puffs of smoke.

A fly enters your office, so you close a small vent by the window before any other insects can take refuge. "Another way of saying this is that motion in our world has *three degrees of freedom*." You write three words in big letters on the blackboard:

DEGREES OF FREEDOM

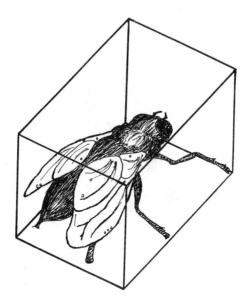

Figure 1.1 A fly in a box is essentially confined to a point. It has zero degrees of freedom and lives its (depressing) life in a 0-D world. (Drawing by Brian Mansfield.)

"Yes, I understand." Sally's hand darts out at the fly and captures it. You never knew she could move so fast. "This fly has three different directions it can travel in the room. Now that it is in my hand, it has zero degrees of freedom. I'm holding my hand very still. The fly can't move. It's just stuck at a point in space. It would be the same if I placed it in a tiny box. Now, if I stick the fly in a *tube* where it can only move back and forth in one direction, then the fly has one degree of freedom" (Fig. 1.1 and 1.2).

"Correct! And if you were to pull off its wings—"

"Sadist."

"—and let it crawl around on a plane, it would have two degrees of freedom. Even if the surface is curved, it still lives in a 2-D world with two degrees of freedom because its movement can be described as combinations of two directions of motion: forward/backward and left/right. Since it can't fly, it can't leave the surface of the paper" (Fig. 1.3).

"The surface is a curved 3-D object, but the fly's motion, confined to the surface, is essentially a 2-D motion."

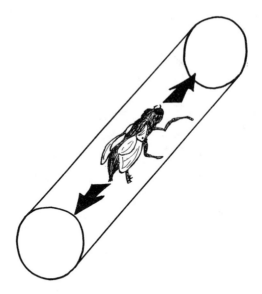

Figure 1.2 A fly in a tube has one degree of freedom and lives in a 1-D world. (Drawing by Brian Mansfield.)

"Sally, you've got it. Likewise, a fly in a tube still lives in a 1-D world, even if you curve the tube into a knot. It still has only one degree of freedom—its motion back and forth. Even an intelligent fly might not realize that the tube was curved."

You trip over a Rubik's cube that you had left on the floor and bump into Sally's chair.

She pushes you away. "Ugh, you made me crush the fly." She tosses it into a wastebasket.

You wave your hand. "It doesn't matter." You pause and return to the discussion. "As I'll show you later, the space we live in may also be curved, just like a twisted tube or a curved piece of paper. However, in terms of our degrees of freedom, we are living in a 3-D world."

Sally holds her fist in front of you, as if about to strike you. "Let me see if I get this. My fist can be described by three numbers: longitude, latitude, and height above sea level. We live in a 3-D world. If we lived in a 4-D space, I would have to specify the location of my fist with a fourth number. In a 4-D world, to find my fist, you could go to the correct longitude, latitude, and height above sea level, and then move into a fourth direction, perpendicular to the rest."

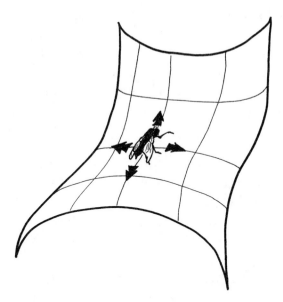

Figure 1.3   A fly that walks on a paper, even a curved piece of paper, has two degrees of freedom and lives in a 2-D world. (Drawing by Brian Mansfield.)

You nod. "Excellent. At each location in my office you could specify different distances in a fourth spatial direction that currently we can't see. It's very hard to imagine such a dimension, just as it would be hard for creatures confined to a plane, and who can only look along the plane, to imagine a 3-D world. Tomorrow I want to do some more reasoning from analogy, because the best way for 3-D creatures to understand the fourth dimension is to imagine how 2-D creatures would understand our world."

Sally taps her hand on your desk. "But how does this explain my encounter at Cherbourg?"

"We'll get to that. By the time your lessons are finished, we're going to see some horrifying stuff. . . ." You reach for a seemingly empty jar on the shelf and hold it in front of Sally's sparkling eyes.

She examines the jar cap, which has been sealed securely to the jar using epoxy. "There's nothing in here."

Your grin widens. "Not yet."

She shakes her head. "You scare me sometimes."

The Science Behind the Science Fiction

Future historians of science may well record that one of the greatest
conceptual revolutions in the twentieth-century science was the real-
ization that hyperspace may be the key to unlock the deepest secrets of
nature and Creation itself.

—Michio Kaku, *Hyperspace*

If we wish to understand the nature of the Universe we have an inner
hidden advantage: we are ourselves little portions of the universe and
so carry the answer within us.

—Jacques Boivin, *The Single Heart Field Theory*

Early Dreams and Fears of a Fourth Dimension

Look at the ceiling of your room. From the corner of the room radiate three
lines, each of which is the meeting place of a pair of walls. Each line is perpendicular to the other two lines. Can you imagine a fourth line that is perpendicular to the three lines? If you are like most people, the answer is a resounding
"no." But this is what mathematics and physics require in setting up a mental
construct involving 4-D space.

What does it mean for objects to exist in a fourth dimension? The scientific
concept of a fourth dimension is essentially a modern idea, dating back to the

1800s. However, the philosopher Immanuel Kant (1724–1804) considered some of the spiritual aspects of a fourth dimension:

> A science of all these possible kinds of space would undoubtedly be the highest enterprise which a finite understanding could undertake in the field of geometry. . . . If it is possible that there could be regions with other dimensions, it is very likely that a God had somewhere brought them into being. Such higher spaces would not belong to our world, but form separate worlds.

Euclid (c. 300 B.C.), a prominent mathematician of Greco-Roman antiquity, understood that a point has no dimension at all. A line has one dimension: length. A plane had two dimensions. A solid had three dimensions. But there he stopped—believing nothing could have four dimensions. The Greek philosopher Aristotle (384–322 B.C.) echoed these beliefs in *On Heaven*:

> The line has magnitude in one way, the plane in two ways, and the solid in three ways, and beyond these there is no other magnitude because the three are all.

Aristotle used the argument of perpendiculars to prove the impossibility of a fourth dimension. First he drew three mutually perpendicular lines, such as you might see in the corner of a cube. He then put forth the challenge to his colleagues to draw a fourth line perpendicular to the first three. Since there was no way to make four mutually perpendicular lines, he reasoned that the fourth dimension is impossible.

It seems that the idea of a fourth dimension sometimes made philosophers and mathematicians a little nervous. John Wallis (1616–1703)—the most famous English mathematician before Isaac Newton and best known for his contributions to calculus's origin—called the fourth dimension a "monster in nature, less possible than a Chimera or Centaure." He wrote, "Length, Breadth, and Thickness, take up the whole of Space. Nor can fansie imagine how there should be a Fourth Local Dimension beyond these three."

Similarly, throughout history, mathematicians have called novel geometrical ideas "pathological" or "monstrous." Physicist Freeman Dyson recognized this for fractals, intricate structures that today have revolutionized mathematics and physics but in the past were treated with trepidation:

A great revolution separates the classical mathematics of the 19th century from the modern mathematics of the 20th. Classical mathematics had its roots in the regular geometrical structures of Euclid and Newton. Modern mathematics began with Cantor's set theory and Peano's space-filling curve. Historically the revolution was forced by the discovery of mathematical structures that did not fit the patterns of Euclid and Newton. These new structures were regarded as "pathological," as a "gallery of monsters," kin to the cubist painting and atonal music that were upsetting established standards of taste in the arts at about the same time. The mathematicians who created the monsters regard them as important in showing that the world of pure mathematics contains a richness of possibilities going far beyond simple structures that they saw in Nature. Twentieth-century mathematics flowered in the belief that it had transcended completely the limitation imposed by its natural origins. But Nature has played a joke on the mathematicians. The 19th-century mathematicians may have been lacking in imagination, but Nature was not. The same pathological structures that the mathematicians invented to break loose from 19th-century naturalism turned out to be inherent in familiar objects all around us. (*Science*, 1978)

Karl Heim—a philosopher, theologian, and author of the 1952 book *Christian Faith and Natural Science*—believes the fourth dimension will remain forever beyond our grasp:

The progress of mathematics and physics impels us to fly away on the wings of the poetic imagination out beyond the frontiers of Euclidean space, and to attempt to conceive of space in which more than three coordinates can stand perpendicularly to one another. But all such endeavors to fly out beyond our frontiers always end with our falling back with singed wings on the ground of our Euclidean three-dimensional space. If we attempt to contemplate the fourth dimension we encounter an insurmountable obstacle, an electrically charged barb-wire fence. . . . We can certainly *calculate* with [higher-dimensional spaces]. But we cannot *conceive* of them. We are confined within the space in which we find ourselves when we enter into our existence, as though in a prison. Two-dimensional beings can *believe* in a third dimension. But they cannot *see* it.

Although philosophers have suggested the implausibility of a fourth dimension, you will see in the following sections that higher dimensions probably provide the basis for the existence of everything in our universe.

Hyperspace and Intrinsic Geometry

> The fact that our universe, like the surface of an apple, is curved in an unseen dimension beyond our spatial comprehension has been experimentally verified. These experiments, performed on the path of light beams, shows that starlight is bent as it moves across the universe.
>
> —Michio Kaku, *Hyperspace*

Imagine alien creatures, shaped like hairy pancakes, wandering along the surface of a large beach ball. The inhabitants are embedded in the surface, like microbes floating in the thin surface of a soap bubble. The aliens call their universe "Zarf." To them, Zarf appears to be flat and two-dimensional partly because Zarf is large compared to their bodies. However, Leonardo, one of their brilliant scientists, comes to believe that Zarf is really finite and curved in something he calls the third dimension. He even invents two new words, "up" and "down," to describe motion in the invisible third dimension. Despite skepticism from his friends, Leonardo travels in what seems like a straight line around his universe and returns to his starting point—thereby proving that his universe is curved in a higher dimension. During Leonardo's long trip, he doesn't feel as if he's curving, although he is curving in a third dimension perpendicular to his two spatial dimension. Leonardo even discovers that there is a shorter route from one place to another. He tunnels through Zarf from point A to point B, thus creating what physicists call a "wormhole." (Traveling from A to B along Zarf's surface requires more time than a journey that penetrates Zarf like a pin through a ball.) Later Leonardo discovers that Zarf is one of many curved worlds floating in three-space. He conjectures that it may one day be possible to travel to these other worlds.

Now suppose that the surface of Zarf were crumpled like a sheet of paper. What would Leonardo and his fellow pancake-shaped aliens think about their world? Despite the crumpling, the Zarfians would conclude that their world was perfectly flat because they lived their lives confined to the crumpled space. Their bodies would be crumpled without their knowing it.

This scenario with curved space is not as zany as it may sound. Georg Bernhard Riemann (1826–1866), the great nineteenth-century geometer, thought constantly on these issues and profoundly affected the development of modern theoretical physics, providing the foundation for the concepts and methods later used in relativity theory. Riemann replaced the 2-D world of Zarf with our 3-D world crumpled in the fourth dimension. It would not be obvious to us that our universe was warped, except that we might feel its effects. Riemann believed that electricity, magnetism, and gravity are all caused by crumpling of our 3-D universe in an unseen fourth dimension. If our space were sufficiently curved like the surface of a sphere, we might be able to determine that parallel lines can meet (just as longitude lines do on a globe), and the sum of angles of a triangle can exceed 180 degrees (as exhibited by triangles drawn on a globe).

Around 300 B.C. Euclid told us that the sum of the three angles in any triangle drawn on a piece of paper is 180 degrees. However, this is true only on a flat piece of paper. On the spherical surface, you can draw a triangle in which *each* of the angles is 90 degrees! (To verify this, look at a globe and lightly trace a line along the equator, then go down a longitude line to the South Pole, and then make a 90-degree turn and go back up another longitude line to the equator. You have formed a triangle in which each angle is 90 degrees.)

Let's return to our 2-D aliens on Zarf. If they measured the sum of the angles in a small triangle, that sum could be quite close to 180 degrees even in a curved universe, but for large triangles the results could be quite different because the curvature of their world would be more apparent. The geometry discovered by the Zarfians would be the *intrinsic* geometry of the surface. This geometry depends only on their measurements made along the surface. In the mid-nineteenth century in our own world, there was considerable interest in non-Euclidean geometries, that is, geometries where parallel lines can intersect. When physicist Hermann von Helmholtz (1821–1894) wrote about this subject, he had readers imagine the difficulty of a 2-D creature moving along a surface as it tried to understand its world's intrinsic geometry without the benefit of a 3-D perspective revealing the world's curvature properties all at once. Bernhard Riemann also introduced intrinsic measurements on abstract spaces and did not require reference to a containing space of higher dimension in which material objects were "curved."

The *extrinsic* geometry of Zarf depends on the way the surface sits in a high-dimensional space. As difficult as it may seem, it is possible for Zarfians to understand their extrinsic geometry just by making measurements along the surface of their universe. In other words, a Zarfian could study the curvature of its universe without ever leaving the universe—just as we can learn about the curvature of our universe, even if we are confined to it. To show that our space is curved, perhaps all we have to do is measure the sums of angles of large triangles and look for sums that are not 180 degrees. Mathematical physicist Carl Friedrich Gauss (1777–1855)—one of the greatest mathematicians of all time—actually attempted this experiment by shining lights along the tops of mountains to form one big triangle. Unfortunately, his experiments were inconclusive because the angle sums were 180 degrees up to the accuracy of the surveying instruments. We still don't know for sure whether parallel lines intersect in our universe, but we do know that light rays should not be used to test ideas on the overall curvature of space because light rays are deflected as they pass nearby massive objects. This means that light bends as it passes a star, thus altering the angle sums for large triangles. However, this bending of starlight also suggests that pockets of our space are curved in an unseen dimension beyond our spatial comprehension. Spatial curvature is also suggested by the planet Mercury's elliptical orbit around the sun that shifts in orientation, or precesses, by a very small amount each year due to the small curvature of space around the sun. Albert Einstein argued that the force of gravity between massive objects is a consequence of the curved space nearby the mass, and that traveling objects merely follow straight lines in this curved space like longitude lines on a globe.[1]

In the 1980s and 1990s various astrophysicists have tried to experimentally determine if our entire universe is curved. For example, some have wondered if our 3-D universe might be curved back on itself in the same way a 2-D surface on a sphere is curved back on itself. We can restate this in the language of the fourth dimension. In the same way that the 2-D surface of the Earth is finite but unbounded (because it is bent in three dimensions into a sphere), many have imagined the 3-D space of our universe as being bent (in some 4-D space) into a 4-D sphere called a hypersphere. Unfortunately, astrophysicists' experimental results contain uncertainties that make it impossible to draw definitive conclusions. The effort continues.

A Loom with Tiny Strings

> In heterotic string theory . . . the right-handed bosons (carrier particles) go counterclockwise around the loop, their vibrations penetrating 22 compacted dimensions. The bosons live in a space of 26 dimensions (including time) of which 6 are the compacted "real" dimensions, 4 are the dimensions of ordinary space-time, and the other 16 are deemed "interior spaces"—mathematical artifacts to make everything work out right.
>
> —Martin Gardner, *The Ambidextrous Universe*

String theory may be more appropriate to departments of mathematics or even schools of divinity. How many angels can dance on the head of a pin? How many dimensions are there in a compacted manifold thirty powers of ten smaller than a pinhead? Will all the young Ph.D.s, after wasting years on string theory, be employable when the string snaps?
—Sheldon Glashow, *Science*

String theory is twenty-first century physics that fell accidentally into the twentieth century.
—Edward Witten, *Science*

Various modern theories of hyperspace suggest that dimensions exist beyond the commonly accepted dimensions of space and time. As alluded to previously, the entire universe may actually exist in a higher-dimensional space. This idea is not science fiction: in fact, hundreds of international physics conferences have been held to explore the consequences of higher dimensions. From an astrophysical perspective, some of the higher-dimensional theories go by such impressive sounding names as Kaluza-Klein theory and supergravity. In Kaluza-Klein theory, light is explained as vibrations in a higher spatial dimension.[2] Among the most recent formulations of these concepts is superstring theory that predicts a universe of ten dimensions—three dimensions of space, one dimension of time, and six more spatial dimensions. In many theories of hyperspace, the laws of nature become simpler and more elegant when expressed with these several extra spatial dimensions.

The basic idea of string theory is that some of the most basic particles, like quarks and fermions (which include electrons, protons, and neutrons), can be modeled by inconceivably tiny, one-dimensional line segments, or strings. Initially, physicists assumed that the strings could be either open or closed into loops, like rubber bands. Now it seems that the most promising approach is to regard them as permanently closed. Although strings may seem to be mathematical abstractions, remember that atoms were once regarded as "unreal" mathematical abstractions that eventually became observables. Currently, strings are so tiny there is no way to "observe" them. Perhaps we will never be able to observe them.[3]

In some string theories, the loops of string move about in ordinary three-space, but they also vibrate in higher spatial dimensions perpendicular to our world. As a simple metaphor, think of a vibrating guitar string whose "notes" correspond to different "typical" particles such as quarks and electrons along with other mysterious particles that exist only in all ten dimensions, such as the hypothetical graviton, which conveys the force of gravity. Think of the universe as the music of a hyperdimensional orchestra. And we may never know if there is a hyperBeethoven guiding the cosmic harmonies.

Whenever I read about string theory, I can't help thinking about the Kabala in Jewish mysticism. Kabala became popular in the twelfth and following centuries. Kabalists believe that much of the Old Testament is in code, and this is why scripture may seem muddled. The earliest known Jewish text on magic and mathematics, *Sefer Yetzira* (Book of Creation), appeared around the fourth century A.D. It explained creation as a process involving ten divine numbers or *sephiroth*. Kabala is based on a complicated number mysticism whereby the primordial One divides itself into ten sephiroth that are mysteriously connected with each other and work together. Twenty-two letters of the Hebrew alphabet are bridges between them (Fig. 1.4).

The sephiroth are ten hypostatized attributes or emanations allowing the infinite to meet the finite. ("Hypostatize" means to make into or treat as a substance—to make an abstract thing a material thing.) According to Kabalists, by studying the ten sephiroth and their interconnections, one can develop the entire divine cosmic structure.

Similarly, physical reality may be the hypostatization of these mathematical constructs called "strings." As I mentioned, strings, the basic building blocks of nature, are not tiny particles but unimaginably small loops and snippets loosely resembling strings—except that strings exist in a strange, 10-D universe. The current version of the theory took shape in the late 1960s. Using hyperspace theory, "matter" is viewed as vibrations that ripple through space and time. From this follows the idea that everything we see, from people to planets, is nothing but vibrations in hyperspace.

In the last few years, theoretical physicists have been using strings to explain all the forces of nature—from atomic to gravitational. Although string theory describes elementary particles as vibrational modes of infinitesimal strings that exist in ten dimensions, many of you may be wondering how such things exist in our 3-D universe with an additional dimension of time. String theorists claim that six of the ten dimensions are "compactified"—tightly curled up (in structures known as Calabi-Yau spaces) so that the extra dimensions are essentially invisible.[4]

As technically advanced as superstring theory sounds, superstring theory could have been developed a long time ago according to string-theory guru Edward Witten,[5] a theoretical physicist at the Institute for Advanced Study in

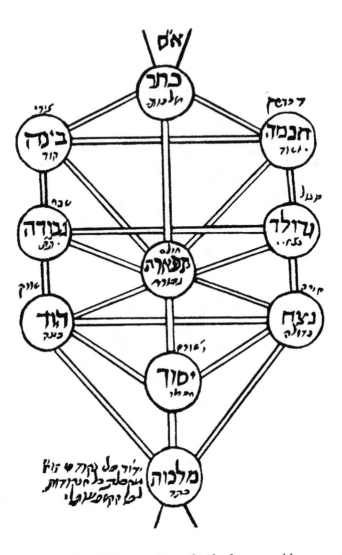

Figure 1.4a The Sephiroth Tree, or Tree of Life, from an old manuscript of the Zohar.

Princeton. For example, he indicates that it is quite likely that other civilizations in the universe discovered superstring theory and then later derived Einstein-like formulations (which in our world predate string theory by more than half a century). Unfortunately for experimentalists, superstrings are so small that they are not likely to ever be detectable by humans. If you consider the ratio of the

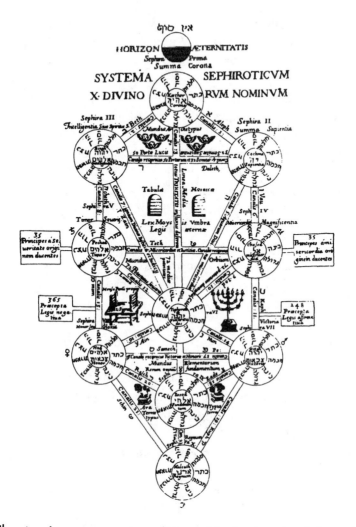

Figure 1.4b Another representation of the sephiroth, the central figure for the Kabala.

size of a proton to the size of the solar system, this is the same ratio that describes the relative size of a superstring to a proton.

John Horgan, an editor at *Scientific American*, recently published an article describing what other researchers have said of Witten and superstrings in ten dimensions. One researcher interviewed exclaimed that in sheer mathematical mind power, Edward Witten exceeds Einstein and has no rival since Newton.

So complex is string theory that when a Nobel Prize–wining physicist was asked to comment on the importance of Witten's work, he said that he could not understand Witten's recent papers; therefore, he could not ascertain how brilliant Witten is![6]

Recently, humanity's attempt to formulate a "theory of everything" includes not only string theory but *membrane theory*, also known as M-theory.[7] In the words of Edward Witten (whom *Life* magazine dubbed the sixth most influential American baby boomer), "M stands for Magic, Mystery, or Membrane, according to taste." In this new theory, life, the universe, and everything may arise from the interplay of membranes, strings, and bubbles in higher dimensions of spacetime. The membranes may take the form of bubbles, be stretched out in two directions like a sheet of rubber, or wrapped so tightly that they resemble a string. The main point to remember about these advanced theories is that modern physicists continue to produce models of matter and the universe requiring extra spatial dimensions.

Hypertime

In this book, I'm interested primarily in a fourth spatial dimension, although various scientists have considered other dimensions, such as time, as a fourth dimension. In this section, I digress and speak for a moment on time and what it would be like to live outside the flow of time. Readers are encouraged to consult my book *Time: A Traveler's Guide* for an extensive treatise on the subject.

Einstein's theory of general relativity describes space and time as a unified 4-D continuum called "spacetime." The 4-D continuum of Einstein's relativity in which three spatial dimensions are combined with one dimension of time is not the same as hyperspace consisting of four spatial coordinates. To best understand this, consider yourself as having three spatial dimensions—height, width, and breadth. You also have the dimension of duration—how long you last. Modern physics views time as an extra dimension; thus, we live in a universe having (at least) three spatial dimensions and one additional dimension of time. Stop and consider some mystical implications of spacetime. Can something exist outside of spacetime? What would it be like to exist outside of spacetime? For example, Thomas Aquinas believed God to be outside of spacetime and thus capable of seeing all of the universe's objects, past and future, in

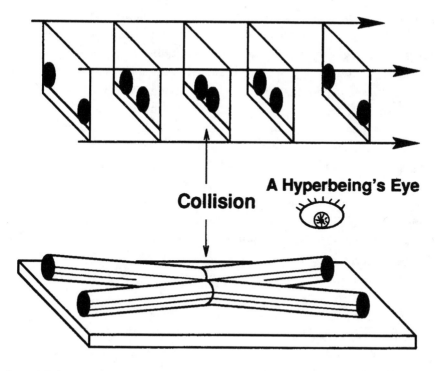

Figure 1.5 An eternitygram for two colliding discs.

one blinding instant. An observer existing outside of time, in a region called "hypertime," can see the past and future all at once.

There are many other examples of beings in literature and myth who live outside of spacetime. Many people living in the Middle Ages believed that angels were nonmaterial intelligences living by a time different from humans, and that God was entirely outside of time. Lord Byron aptly describes these ideas in the first act of his play *Cain, A Mystery*, where the fallen angel Lucifer says:

> With us acts are exempt from time, and we
> Can crowd eternity into an hour,
> Or stretch an hour into eternity.
> We breathe not by a mortal measurement—
> But that's a mystery.

A direct analogy involves an illustration of an "eternitygram" representing two discs rolling toward one another, colliding, and rebounding. Figure 1.5 shows two spatial dimensions along with the additional dimension of time. You can think of successive instants in time as stacks of movie frames that form a 3-D picture of hypertime in the eternitygram. Figure 1.5 is a "timeless" picture of colliding discs in eternity, an eternity in which all instants of time lie frozen like musical notes on a musical score. Eternitygrams are timeless. Hyperbeings looking at the discs in this chunk of spacetime would see past, present, and future all at once. What kind of relationship with humans could a creature (or God) have who lives completely outside of time? How could they relate to us in our changing world? One of my favorite modern examples of God's living outside of time is described in Anne Rice's novel *Memnoch the Devil*. At one point, Lestat, Anne Rice's protagonist, says, "I saw as God sees, and I saw as if Forever and in All Directions." Lestat looks over a balustrade in Heaven to see the entire history of our world:

> . . . the world as I had never seen it in all its ages, with all its secrets of the past revealed. I had only to rush to the railing and I could peer down into the time of Eden or Ancient Mesopotamia, or a moment when Roman legions had marched through the woods of my earthly home. I would see the great eruption of Vesuvius spill its horrid deadly ash down upon the ancient living city of Pompeii. Everything there to be known and finally comprehended, all questions settled, the smell of another time, the taste of it. . . .

If all our movements through time were somehow fixed like tunnels in the ice of spacetime (as in the eternitygram in Fig. 1.5), and all that "moved" was our perception shifting through the ice as time "passes," we would still see a complex dance of movements even though nothing was actually moving. Perhaps an alien would see this differently. In some sense, all our motions may be considered fixed in the geometry of spacetime, with all movement and change being an illusion resulting from our changing psychological perception of the moment "now." Some mystics have suggested that spacetime is like a novel being "read" by the soul—the "soul" being a kind of eye or observer that stands outside of spacetime, slowly gazing along the time axis.

Note 8 describes the metamorphosis of time into a spatial dimension during the early evolution of our universe. At the point when time loses its time-like character, the universe is in the realm of what physicists call "imaginary time."

> *Mulder:* Didn't you ever want to be an astronaut when you were growing up?
> *Scully:* I must have missed that phase.
> —"Space," *The X-Files*

two

What I have seen cannot be described. . . . Door after door opened upon my heart, and my soul became acquainted with thoughts not of this world. . . . It seemed as if a hundred thousand seas, vast and sunlit, billowed upon that Blessed Face. What happened then, I do not know. My last word to you is this: never ask for anything like this and be contented with what is given unto you.

—Aqa Siyyid Ismail-i-Zavarii (Dhabih),
as quoted in H. M. Balyuzi's *Baha'u'llah: The King of Glory*

Imagine an ant finding its path suddenly blocked by a discarded Styrofoam cup. Even if the ant is intelligent, can it hope to understand what the cup is for and where it came from?

—Charles Platt,
When You Can Live Twice as Long, What Will You Do?

What's a nice Jewish boy like you doing in the sixth dimension?

—Old Yiddish man in *Forbidden Zone*

the divinity of higher dimensions

FBI Headquarters, Washington, D.C., 3:00 P.M.

Sometimes you feel like you are not part of this world. People often stare at you on the street and when you wander the FBI halls late at night. Even the custodial staff seems to follow you as if you are something odd or alien. You consider the situation with an amusement and aloofness that angers those closest to you, particularly Sally.

"What's that?" Sally asks.

You look at your hand. "This is my hand."

Sally rolls her eyes. "Not that. I mean the book in your hand."

"*Flatland.*" You dust off an old book and flip through its pages. "It was published in 1884 by a Victorian schoolmaster named Edwin Abbott Abbott. It's about the life of a square in a 2-D world called Flatland. When he tells his people about the third dimension, they put him in jail."

Sally's eyes dart around your cluttered office and focus on a large photo of former FBI Chief J. Edgar Hoover. A thin trail of strawberry incense arises from a burning stick on your desk, and there is a twangy musical sound coming from some nearby speakers. You are now playing a CD of Bismilla Kahn and Ravi Shankar.

Sally's eyes focus on you. "You said the square is put in jail? Are you hinting that we're going to get in trouble by exploring the fourth dimension?"

"I'm already being followed by secretive, cigarette-smoking men in black hats."

"That's absurd."

"Sally, listen up. I want to talk about *Flatland* because by understanding the square's difficulty in visualizing the third dimension, we'll be better able to deal with your problem with the fourth dimension."

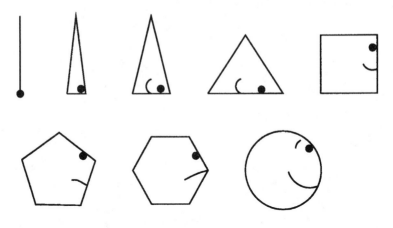

Figure 2.1 Eight inhabitants of *Flatland*: woman, soldier, workman, merchant, professional man, gentleman, nobleman, high priest.

Sally places her hands on her hips. "My problem?"

"Moreover, if we can understand the square's experiences, we'll have a perfect metaphor for spiritual enlightenment, God, and all manner of mystic experience."

Sally shakes her head. "Now that's something you're going to have to *prove* to me."

"Flatland is a plane inhabited by creatures sliding around in the plane's surface. Their society is based on a caste system whereby a male's status depends on the number of sides of his body. Women are mere line segments, soldiers and workmen are isosceles triangles, merchants are equilateral triangles, professional men are squares, gentleman are regular pentagons, and noblemen are regular polygons with six or a greater number of sides. Their high priests have so many sides that they are indistinguishable from circles" (Fig. 2.1).

Sally paces around your office as she gazes at various geometrical models hanging on strings from the ceiling. "Why are women represented by lowly lines?"

You shrug. "He was satirizing the stodgy, insensitive society of the Victorians. In the book, irregulars (cripples) are euthenized, and women have no rights. They're mere lines, infinitely less respected than the priestly circles with an 'infinite' number of edges."

Sally stares as the *Flatland* figure showing the various castes of individuals.

You open *Flatland*, searching for a particular page. "Remember, in the nineteenth century women were considered much less able than men. I think Abbott was trying to show some of the society's prejudices, because later in the book a sphere visits Flatland and says: 'It is not for me to classify human faculties according to merit. Yet many of the best and wisest of Spaceland think more of your despised Straight Lines than of your belauded Circles.'"

Sally nods.

"Here, let me read a passage to you in which the square is speaking":

I call our world Flatland, not because we call it so, but to make its nature clearer to you, my happy readers who are privileged to live in Space. . . . Imagine a vast sheet of paper on which Lines, Triangles, Squares, Pentagons, Hexagons, and other figures, instead of remaining in their places, move freely about, on or in the surface, but without the power of rising above or sinking below it, very much like shadows—and you will then have a pretty correct notion of my country and countrymen. Alas, a few years ago, I should have said "my universe" but now my mind has been opened to higher view of things.

Outside the window of your office, dark clouds race against the city's skyline. There's a storm due later tonight. You and Sally look up as a honk-honk-honk of Canadian geese ride the winds to their winter haven along the Tangiers Sound. Just overnight, hundreds of ospreys and other birds were in Washington, D.C., and even on the roof of the FBI building. You don't mind the duck calls occasionally waking you at night. In fact, you love the finches and chickadees that flutter about your feeders. The vultures, however, give you the creeps. Sometimes they perch on the dead tree in your backyard. At night, they remind you of dark vampires.

Sally looks back at you. "If all the creatures of Flatland move around in a plane, and only see things in the plane, how can they tell one another apart? Wouldn't they only see each other's sides?"

"Excellent question. Their atmosphere is hazy and attenuates light. Those parts of the creature's sides that are farther from the viewer's eye get dimmer. Close parts are brighter and clearer. Don't forget that our own retina is a 2-D surface, yet we can distinguish all sorts of objects; for example, we can tell the difference between a sphere and a disc simply by their shading."

Sally nods. "I can think of another way Flatlanders can distinguish objects. They can tell when one object is front of another, and this also provides a visual depth cue." Sally places one of her hands in front of the other. "The same is true in our world. I see one hand in front of my other hand. I don't assume that one hand is magically passing through the other hand. I recognize that there is a third dimension of space, and that one hand is closer than the other in this dimension."

"Right. Just as we build up a mental image of our 3-D world, the Flatlanders have many ways to understand and survive in their 2-D world." You pause. "But Abbott's book doesn't only discuss 2-D worlds. It also discusses the square's visions of a 1-D world called Lineland. The square says":

I saw before me a vast multitude of small Straight Lines (which I naturally assumed to be Women) interspersed with other Beings still smaller and of the nature of lustrous points—all moving to and fro in one and the same Straight Line, and as nearly as I could judge, with the same velocity.

A noise of confused multitudinous chirping or twittering issued from them at intervals as long as they were moving, but sometimes they ceased from motion, and then all was silence.

Approaching one of the largest of what I thought to be Women, I accosted her, but received no answer. A second and a third appeal on my part were equally ineffectual. Losing patience at what appeared to me intolerable rudeness, I brought my mouth into a position full in front of her mouth so as to intercept her motion, and loudly repeated my question, "Woman, what signifies this concourse, and this strange and confused chirping, and this monotonous motion to and fro in one and the same Straight Line?"

"I am no Woman," replied the small Line. "I am the Monarch of the world."

You hand Abbott's drawing of Lineland to Sally. The drawing shows women as mere dots and the rest of the inhabitants as lines (Fig. 2.2).

Sally studies the diagram. "It seems to me that the only parts the Linelanders can see of one another are single points."

"Right."

"How do the Linelanders tell where other inhabitants are located in their world?"

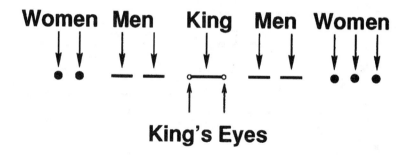

Figure 2.2 Edwin Abbott Abbott's conception of Lineland with women as dots and the King as a line at center. The King can see only points.

"According to Abbott, they can do it by hearing. They can also determine a male's body length because men have two voices, a bass voice produced at one end and a treble voice at the other. By listening for the differences in the sounds' arrival times, they get a feel for how long the body is."

As you study Sally, you marvel at how she always has a refined, sophisticated look about her. Her shoulder-length hair makes her look like a younger version of Princess Diana, years before the Princess died. Sally, though, was hardly a world traveler like Diana. Except for college, Sally never ventured far from Washington. Instead Sally preferred to travel the local streets, practicing her law-enforcement methods as she walked, testing her skills as all manner of riffraff accosted her. Rarely did she have to flash her badge or pull her gun.

You are in stark contrast to Sally. You came from a family of good old boys with reddened faces and calloused hands who harvested the Chesapeake waters for as long as anyone could remember. Although you distinguished yourself in college, especially in physics and military history, you returned to your roots, preferring a life away from the academicians and endless power struggles. On the other hand, the FBI gave you a chance to test your crazy theories in the real world. To make a little extra money, your photographs of the Potomac River and its animals are often welcome at *Smithsonian* magazine.

Sally comes closer. "When the square is off to the side of Lineland, the King can't see him. I bet the King is frustrated!"

"Yes, he's a cranky fellow. The square tries to tell the King about this mysterious second dimension. To help the King visualize the second

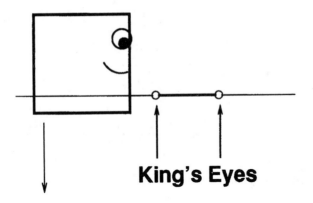

King's Eyes

Figure 2.3   A square moves through Lineland while the King observes.

dimension, the square gradually moves perpendicular to the line." You draw on your office's blackboard (Fig. 2.3). "As the square moves through the space of Lineland, he becomes apparent to the King as a segment that appears out of nowhere, stays for a minute, and then disappears in a flash. Later I'll tell you about how the same sort of appearance and disappearance would happen as a 4-D being moved through our space."

Sally nods. "Let me see if I can visualize a trip to 2-D Flatland. If I were to plunge through their universe it might be like wading through an endless lake. The plane of the lake is their world. They can only move in the lake's surface. If the plane of Flatland cuts me at my waist, the Flatlanders would only see two small blobs (my arms) and a central blob (my body). Think how odd I would appear to them!"

"Pretty scary."

"It would be like a miracle when I suddenly appeared in their midst. As I moved around, a second miracle would be the surprising changeability of my form. If I could swim horizontally through their world, they'd see a human-like outline, but if I stood horizontally, I'd be totally incomprehensible to them" (Fig. 2.4).

"Right, and if the Flatlanders tried to surround you to keep you in one place, you could escape by moving perpendicularly into the third dimension. In their eyes, you would be a God."

From beneath a large pile of books on your desk, you bring out one with the story "Letter to My Fellow-Prisoners in the Fortress of Schlusselburg." It was written by N. A. Morosoff in 1891 and describes the pow-

Figure 2.4   How Flatlanders would see Sally. (Drawing by Brian Mansfield.)

ers of 3-D beings as observed by 2-D creatures. You read aloud from the book:

> If, desirous of keeping you in one place, they surrounded you on all sides, you can step over them and find yourself free from them in a way quite inconceivable to them. In their eyes, you would be an all-powerful being—an inhabitant of a higher world, similar to those supernatural beings about whom theologians and metaphysicians tell us.

Sally looks out the window. "Two-dimensional beings wouldn't have an ounce of privacy if 3-D beings were nearby."

You nod. "In a later section of *Flatland*, the square and his wife are comfortably in their bedroom with the door locked, sharing an intimate moment. Suddenly they hear a voice from a sphere hovering inches above their 2-D space. As the sphere descends into Flatland, a circle appears in the square's room. The circle is the cross section of the sphere. Notice

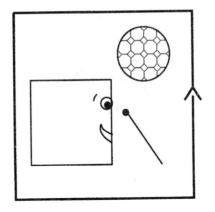

Figure 2.5 A square and his wife in Flatland, trying to get some privacy in a locked room, when suddenly, seemingly from out of nowhere, a sphere penetrates their space.

that the sphere enters the square's room without having to open any doors" (Fig. 2.5).

Sally smiles. "That was rude of the sphere."

"Maybe not. The sphere has come to teach the square about the third dimension."

You reach for Sally's hand. "There would be no privacy for us either. If a 4-D being wanted to, it could come into our room during a romantic interlude" (Fig. 2.6).

Sally pushes your hand away. "Sir, our relationship is strictly professional."

You nod as your gaze shifts to the wall where you have a photo of former president Bill Clinton proudly saluting the American flag. You sigh and return to your lecture. "A 4-D man could reach into any sealed rooms or containers. This being would never have to touch the walls of the container just as we could reach inside a hollow, square vault in Flatland and remove an object. The 4-D being could steal money from a safe without trying to open the door; in fact, the safe might seem like a box with no top or bottom. A 4-D surgeon could reach inside your body without breaking your skin, steal your brain without cracking your skull (Fig. 2.7). Just as we could see every side of a square simultaneously as well as the insides of a square, a 4-D being could see all of our sides simultaneously; see the insides and outsides of your lungs and run his

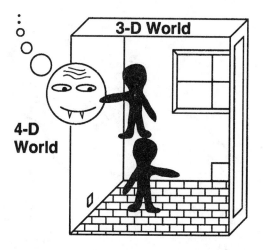

Figure 2.6 A man and his wife, in our 3-D world, trying to get some privacy in a locked room, when suddenly, seemingly from out of nowhere, a grinning hypersphere penetrates their space.

fingers lingeringly along the folds and convolutions of your brain. The being could drink wine from a bottle of 1787 Château Lafite claret, the most expensive wine in the world, without popping the cork." You catch your breath, hoping that Sally has found your lecture impressive and eloquent. "Think how easy it would be to locate a blockage in a maze of pipes. Or—" You pause dramatically. "Or imagine a 4-D vampire that could suck your blood or impregnate you without your even seeing him."

Sally puts up her hand. "Stop! I get the picture." She pauses as you both walk over to the window and watch the swollen sun collapse into a notch between the columns of the Capitol building. Beautiful! The winter sun burns the columns to a golden brown, and torpid yellow-tinted clouds hang over the alabaster reflections. Soon it will be dark, and you should head home.

Sally turns to you. "I think I understand what you're saying. If I had the muscles to reach my arm up into the fourth dimension, I could reach 'through' a solid wall and take priceless pearls or a valuable Ming vase from a sealed case in a museum. My arm would be completely solid the whole time, but the theft would be accomplished by moving my arm up through the fourth dimension. I would lift the vase out of the case by

Figure 2.7 A surgeon with 4-D powers could perform "closed-heart surgery" on a 3-D person; that is, the heart could be removed without even pricking the skin. (Drawing by Brian Mansfield.)

moving it up into the fourth dimension to get 'around' the wall" (Fig. 2.8).

You nod. "Let's return to the *Flatland* story. Do you want to know what happens to the square when the sphere starts talking to him in the bedroom?"

"Sure."

"Well, naturally the square doesn't believe the sphere is anything more than a circle that can change size. The square believes that the sphere is just an ordinary 2-D creature like himself. However, the sphere objects to this simple characterization":

I am not a plane Figure, but a Solid. You call me a Circle, but in reality I am not a Circle, but an indefinite number of Circles, of size varying from a Point to a Circle of thirteen inches in diameter, one placed on the top of the other. When I cut through your plane as I am now doing, I make in your plane a section which you, very

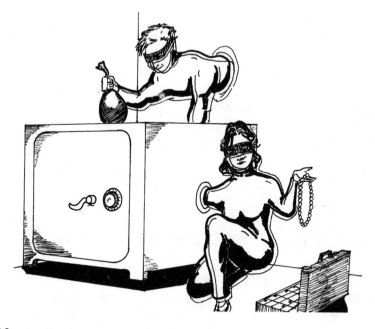

Figure 2.8 Stealing is easy in another dimension. (Drawing by Brian Mansfield.)

rightly, call a Circle. For even a Sphere—which is my proper name in my own country—if he manifest himself at all to an inhabitant of Flatland—must needs manifest himself as a Circle.

Do you not remember—for I, who see all things, discerned last night the phantasmal vision of Lineland written upon your brain—do you not remember, I say, how when you entered the realm of Lineland, you were compelled to manifest yourself to the King, not as a Square, but as a Line, because the Linear Realm had not Dimensions enough to represent the whole of you, but only a slice or section of you? In precisely the same way, your country of Two Dimensions is not spacious enough to represent me, a being of Three, but can only exhibit a slice or section of me, which you call a Circle.

You walk over to a tub of water in the corner of your office. You used to keep piranha in it, but years ago all the fish died. It was just too difficult getting sufficient food for them. You allowed them to cannibalize themselves until there was only one fish left, and, alas, he finally succumbed by cannibalizing his own body, despite the occasional Burger King scrap you threw into the tank and the infrequent dead rat the cleaning staff left outside your door for food.

Now you keep the tank in your room to help visualize intersections of 3-D objects with 2-D worlds. You reach into the water and withdraw a ball. "Here's what a sphere looks like to an inhabitant of Flatland. The surface of the water is a metaphor for Flatland." You push the ball down so that it just touches the water at a point. You push further and the point turns into a circle. The circle enlarges until it reaches a maximum size and then shrinks back down to a point as you push the sphere under the water. The point disappears.

You turn to Sally. "Imagine how difficult it would be for an inhabitant of Flatland to think of all these different circles as together forming a 3-D object" (Fig. 2.9).

Sally walks over to the tub of water. "It's as if the square had blinders on that didn't permit him to look up or down, just straight ahead. If an ant floated on the surface of the water and could only see along the surface, it would see the sphere as just a circle growing and shrinking in size." She pauses. "Sir, could we have blinders on? Could our brains blind us from looking 'up' and 'down' in the fourth dimension where God, demons, angels, and all sorts of beings could be listening to every word, watching our every action, just inches away from us in another direction?" (Fig. 2.10).

"Ooh, Sally, I thought you were supposed to be the skeptic." You begin to hum the eerie, repetitive theme of the *Twilight Zone*. "Sally, I'm beginning to like you."

She smiles. "I'm just speculating out loud."

You slam your fists together and she gasps.

"Sally, what would you see right now if a 4-D hypersphere were to pass through the space right in front of your eyes?" You wiggle your fingers in front of her face and she backs up.

"Put your hand down," she says. "Reasoning from your analogy with a sphere penetrating Flatland, we would first see a point, then a small sphere, then a big sphere. Eventually the sphere would shrink down to point and disappear from our world. It would be like inflating and deflating a balloon."

"Correct, all balloons are beings from the fourth dimension."

"What!"

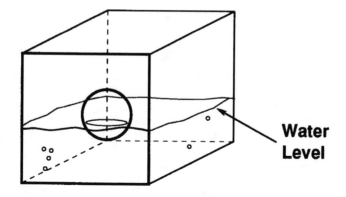

Figure 2.9 A ball being pushed through the surface of water is a metaphor for a sphere moving through Flatland.

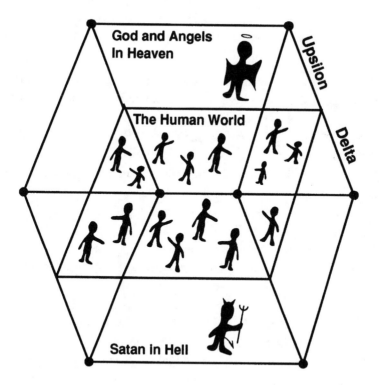

Figure 2.10 God, angels, and demons inhabiting the fourth dimension, just a slight movement away from our 3-D world in a direction we can barely perceive but in which we cannot move.

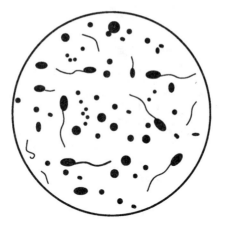

Figure 2.11 Identically shaped sperm from a 4-D being. As the hyperellipsoidal heads intersect our 3-D world, they would first appear as points, then ellipsoids that changed in size, and then points again as they left our space. Depending on their intersection with our world, at times they would resemble spheres. We might only see their heads or tails.

"Just kidding, but you get the point. A sphere is a 3-D stack of circles with different radii. A cylinder is a 3-D stack of circles of the same radius. A hypersphere is a 4-D stack of spheres with different radii. A hypercylinder is a 4-D stack of spheres of the same radii."

"I can't visualize how to stack objects in a fourth dimension."

"Sally, it's very difficult. Some scientists use computer graphics to help visualize 3-D cross sections of rotating 4-D objects." You pause. "Do you recall I told you how easy it would be for a 4-D being to impregnate a woman without being seen?"

"That was too weird to contemplate."

"Well, contemplate this. What do you think 4-D sperm would look like?"

Sally takes a deep breath. "Instead of having ellipsoidal heads, 4-D sperm might have hyperellipsoidal heads. As the heads intersect our 3-D world, they would first appear as points, then ellipsoids that changed in size. Just before it disappeared, a small blob would remain for some time as the tail passed through. A 4-D man could, in principle, inseminate a 3-D woman without her even seeing him"[1] (Fig. 2.11).

Outside it is beginning to snow. A few large flakes intersect the plane of your window and disappear. You are quiet as you watch the headlights

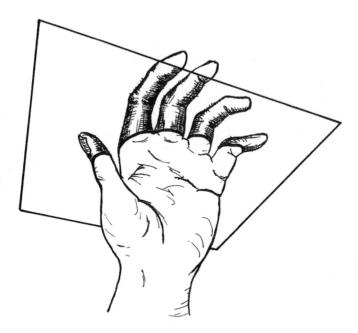

Figure 2.12 A 4-D "God" thrusts His hand into three-space. Is there a way to lock a 4-D being in our space by thrusting a knife through the hand? (Drawing by Brian Mansfield.)

of cars, the bustle of pedestrians, and a man dressed in a Santa Claus out-fit rushing by.

You turn back to Sally. "Imagine a 4-D God thrusting His hand into our world. We'd see His cross section. Some people would certainly be fearful. Someone too bold might stab God's hand with a knife" (Fig. 2.12).

Sally gazes out the window. "Could we harm a 4-D being whose hand came into our world?"

"Let's think about this in lower dimensions that are easier for us to visualize. If Flatlanders are truly two-dimensional, that means they have no thickness. If this is so, they will be as immaterial as shadows. If they stabbed your 3-D hand with the sharp point of a triangle, you would not get hurt, nor would you have any problem lifting your hand out of their space. I'm not sure Flatlanders could cut through your skin. However, another model is to think of Flatland as a rubber sheet in which the inhabitants have a very slight thickness. In the second edition of *Flatland*, Edwin Abbott Abbott suggests that all inhabitants have a slight height, but since they are all the *same* height, none of them realizes there is this third dimension. They don't have the power to move in this dimension."

"If Flatlanders had some thickness, they could cut into your hand like a knife. If the inhabitants were miles long, they could have enough weight to trap your hand."

Sally looks away from the window. "Right. Similarly, someone might be able to knife a 4-D God and trap Him forever in our 3-D world."

A shiver runs up your spine because for an instant you see a vision of Jesus pinned to the cross. Could some of the alleged miracles in the past be the result of beings from a higher dimension? Moses, Jesus, Mohammad, Buddha, and Baha'u'llah—did they have access to hyperspace? Could they lift their eyes, remove their blinders, and peer into other worlds? You recall psychologist William James' quote:

> Our normal waking consciousness is but one special type of consciousness, whilst all about it, parted from it by the filmiest of screens, there lie potential forms of consciousness entirely different. No account of the universe in its totality can be final which leaves these other forms of consciousness quite disregarded. They may determine attitudes though they cannot furnish formulas, and open a region though they fail to give a map.

And then you remember legends of Baha'u'llah, the great Persian religious prophet, permitting a human to gaze into the afterlife and dimensions beyond our comprehension, and the human going ecstatically mad as a result. Similarly, in Flatland, a sphere grabs the square and lifts him up into space, producing a sensory symphony that is both beautiful and horrifying. The square recalls:

> An unspeakable horror seized me. There was a darkness; then a dizzy, sickening sensation of sight that was not like seeing; I saw a Line that was no Line; Space that was not Space; I was myself, and not myself. When I could find voice, I shrieked aloud in agony, "Either this is madness or it is Hell." "It is neither," calmly replied the voice of the Sphere, "it is Knowledge; it is Three Dimensions; Open your eye once again and try to look steadily."

Sally taps you on the shoulder. "You're quiet. Penny for your thoughts?"

Figure 2.13   Sally lifts a 2-D human into the third dimension. If he were truly two-dimensional, with no thickness, Sally might lift up only his skin, leaving some of his guts behind. (Rudy Rucker in *The Fourth Dimension* discusses this scenario among many others relating to creatures interacting between dimensions. Drawing by Brian Mansfield.)

"I'm thinking about how it would feel to be lifted into a higher dimension."

Sally draws a picture of a 2-D human on a plane.

"Cute picture. What's his name?"

She shrugs. "I call him Tdh. It stands for 2-D human. What if I were to lift up Mr. Tdh from his 2-D universe. Would I kill him? He would survive if he had some thin membranes sealing off his upper and lower faces against the third dimension, otherwise when I pulled him, I might get only his skin!" (Fig. 2.13).

You nod. "However, if we think of your 2-D man as residing in a plane with some thickness, we could lift him off without much harm."

Sally draws a digestive system, with a tube extending from the man's mouth to his anus (Fig 2.14). "It seems to me that 2-D humans can't have a complete digestive system running from one orifice to another because this would separate the being into two pieces that fall apart. Maybe only primitive creatures would evolve on Flatland, like planaria flatworms or hydra that only have one opening for the digestive system. They eat food and expel wastes from the same opening" (Fig. 2.15).

Figure 2.14 Can a 2-D human have a tubular digestive system without falling apart into two pieces? (Drawing by Clay Fried.)

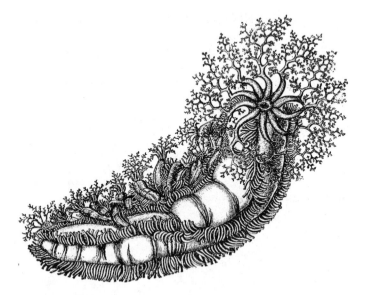

Figure 2.15 Perhaps in a 2-D world, only creatures with primitive digestive systems would evolve—like Earthly planaria flatworms or hydra that only have one opening in the digestive system. They eat food and expel 2-D wastes from the same opening.

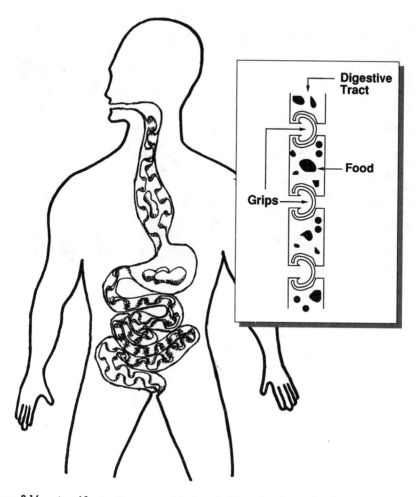

Figure 2.16 A self-gripping gut with interlocking knobs or hooks. (Drawing by Brian Mansfield.)

"Sally, you're forgetting something. One way to keep poor Mr. Tdh from falling apart would be for natural selection to evolve a 'self-gripping gut.' Each side of the tube would have interlocking projections, almost like a zipper. The zipper- or Velcro-like structure is open at the mouth-end when he eats. As the food passes through the body, the zipper closes behind the food and opens ahead of it, thereby keeping his body intact"[2] (Fig. 2.16).

The snowflakes are falling faster now. If the snow were falling from the fourth dimension, it would be difficult to tell because the flakes are mere points winking out of existence and flying so fast. You enjoy watching

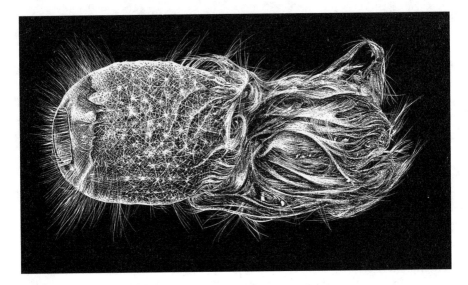

Figure 2.17 4-D guinea pig.

the American flags outside your window bellying in the wind like mata-
dors' capes. Overhead, a few birds fly. For a moment their cries remind
you of the happy screams of children.

You follow Sally's eyes that have wandered to the windowsill on which
is perched a bronze bust of astronaut Neil Armstrong. On the adjacent
wall is a picture of Robert Kennedy, Herbert Hoover, and Nikita
Khrushchev boarding a flying saucer.

Sally's eyes are wide as she steps closer to the pictures. "Am I crazy, or
does your office decor seem a little out of place?"

"Never mind that. Turn on the TV. Let's get a weather report."

Sally switches on a small TV buried under a stack of mathematics
texts. Larry King, a CNN talk-show host, is shouting some late-breaking
news into his large microphone. It's something about floating blobs of
flesh appearing in the White House. There is a rumor that it is a new
Russian spying device.

A shiver runs up your spine. "Hold on! Turn up the volume. We've got
to investigate."

"Could it be something from the fourth dimension?"

You grab your black overcoat. It must be the Omegamorphs, the 4-D
creatures that have recently intruded into your world. "We've got to get
out of here and head to the White House."

Suddenly the jar on your shelf becomes alive with pulsating, hairy creatures.

Sally jumps back. "Holy mackerel! What are those?"

"Those are guinea pigs from the fourth dimension"(Fig. 2.17).

You wink at Sally. Then they wink at Sally.

Sally is silent.

The Science Behind the Science Fiction

Can you imagine standing in the center of a sphere and seeing all the abdominal organs around you at once? There above my head were the coils of the small intestine. To the right was the cecum with the spectacles beside it, to my left the sigmoid and the muscles attached to the ilium, and beneath my feet the peritoneum of the anterior abdominal wall. But I was terribly dizzy for some reason; I could not stand it very long, much as I should have liked to remain inside of him for a while."

—Miles J. Breuer, "The Appendix and the Spectacles"

> Tear apart the veils,
> Bring forth the means,
> Breathe the breath of love.
> This I say, and saying it, burn do I.
> —Aqa Siyyid Ismail-i-Zavarii (Dhabih), as quoted in
> H. M. Balyuzi's *Baha'u'llah: The King of Glory*

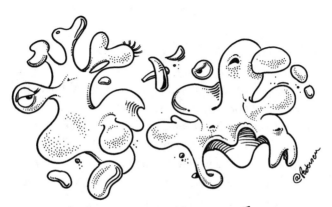

BEINGS FROM THE FOURTH DIMENSION?

Removing Egg Yolks

If 4-D creatures existed, they would have the awesome powers discussed in this chapter. Four-dimensional beings wouldn't have to open doors to get into our homes. They could simply enter "through" the walls.[3] If they were traveling on a long journey over the Himalayas, they could make the trip easier by detouring around the jagged peaks and stepping through them. (Imagine a 2-D analogy with jagged lines printed on paper. You could bypass them by stepping into the third dimension while your 2-D friend has to carefully navigate them like a maze.)

If they were hungry, 4-D hyperparasites (or carnivores) might reach into your stomach or refrigerator for some food, without opening either. If you suddenly had 4-D powers, you would never have to worry about being locked outside your car. You could step "upward" over the door and back inside the car.

Think of the pranks you could play. You could remove the egg yolk from an egg without cracking it, or the inside of a banana without removing the peel. You could be the ultimate surgeon, removing tumors without cutting the skin, thereby reducing the risk of bacterial infections.[4]

I shudder to imagine what would happen if these powers got into the wrong hands. Imagine scenarios more horrifying than dreamed of in Stephen King's or Anne Rice's worst nightmares: brain-snatchers, blood-suckers, and midnight muggings by phantom prowlers materializing out of the air.[5]

A 4-D being would appear as multiple 3-D objects in 3-D space. Thus, you could have several disconnected 3-D blobs that were all part of the same being sharing the same sensations. To our eyes, they could appear to merge and diverge seemingly at random (Fig. 2.18).

Flatland

Over a century ago, Edwin Abbott Abbott—clergyman and the headmaster of a school in Victorian England—wrote a wonderful book describing interactions between creatures of different dimensions. One of Abbott's purposes was to encourage readers to open their minds to new ways of perceiving. *Flatland* described an entire race of 2-D creatures, living in a flat plane, totally unaware of the existence of a higher dimension all around them. If we were able to look down on a 2-D universe, we would be able to see inside every structure at once in the same way a bird could gaze at the maze in Figure 2.19 and see everything inside, while someone wandering the 2-D floors would have no knowledge of the maze's structure.

Figure 2.18 A 4-D creature (Omegamorph) can appear in several different 3-D locations at once. (Note the feet on this creature intersecting our 3-D world represented as a plane. Drawing by Sean Henry.)

We can use another analogy to understand our lack of privacy when observed by 4-D beings. Consider an ordinary ant farm in which ants wander a (mostly) 2-D world between panes of glass. Like Flatlanders, the ants would have great difficulty hiding from us, except perhaps by pressing against walls and hoping we would not see them.

If there really were a Flatland universe draped along a wall in your home, the Flatlanders could go through their entire lives unaware that you were poised inches above their planar world recording all the events of their lives. Figure 2.20 emphasizes this point by illustrating some fanciful Flatland homes designed from letters of the alphabet. Notice that we can see all parts of the houses simultaneously as well as everything inside their rooms and closets. However, to a Flatlander, these luxurious homes provide perfect privacy from their neighbors. Once a Flatlander closed the door, he or she would be safe inside, even though there is no roof overhead. To climb over the lines would mean getting out of the plane into a third dimension, and no inhabitants of a 2-D world would have any better idea of how to do this than we know how to escape from a room by vaulting into a fourth dimension.

Because of your seeming omniscience, Flatlanders would think of you as a God. Your powers would awe them. For example, if you wanted to jail a criminal, you could simply draw a circle around him. However, it might be possible for you to lift the criminal up and deposit him elsewhere in Flatland. This act

Figure 2.19 Taking advantage of three-space, a bird can look down at a maze and see its entire structure, while people wandering the 2-D floor are unaware of the maze's structure.

would appear miraculous to a Flatlander who would not even have the word "up" in his vocabulary.

Could there be even higher-dimensional Gods with greater degrees of omniscience? Could there be universes of five or six or seven dimensions, each one able to look down on its dimensionally impoverished predecessor whose inhabitants couldn't hide from the prying eyes of the next-higher beings?

I like to imagine a universe where the dimensionally impoverished carry picket signs with such words as

AFFIRMATIVE ACTION FOR THE DIMENSIONALLY CHALLENGED

Who would carry such signs? Three-dimensional creatures parading before the all-seeing eyes of their 4-D brethren? 4-D people in front of 5-D people? Today's government mandates access ramps so that handicapped people can enter public buildings. Similarly, will governments of our far future mandate dimensional conveniences and portals?

Perhaps even more weird than these higher dimensions is the 0-D world we might call "Pointland." It possesses neither length nor breadth nor height. This

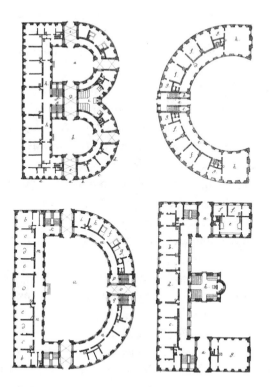

Figure 2.20 We can see all parts of a Flatland house simultaneously and look inside any room. (This is not a "blueprint" but an actual house in Flatland.) Similarly, a 4-D being could look at all parts of our homes simultaneously. How could you hide?

point is self-contained. Karl Heim enigmatically describes Pointland in *Christian Faith and Natural Science*: "There is nothing which is not within it."[6]

The Women of Flatland

In Abbott's Flatland, society is rigidly stratified. Soldiers are isosceles triangles with very short bases and sharp points for attacking enemies. The middle class is composed of equilateral triangles. Professional men are squares and pentagons. The upper classes start as hexagons; the number of their sides increases with their social status until the figures are indistinguishable from circles. The circles are the priests and administrators of Flatland. At the lowest level are the women: straight lines with an eye at one end, like a needle. There is a visible glow from a woman's eye, but none from her other end, so that she can make herself invisible simply by turning her back. Her sharp posterior can be danger-ous. To avoid accidents, women are legally required to always keep themselves visible by perpetually wobbling their rear ends. Abbott writes:

> If our highly pointed Triangles of the Soldier class are formidable, it may be readily inferred that far more formidable are our Women. For, if a Soldier is a wedge, a Woman is a needle; being, so to speak, *all* point, at least at the two extremities. Add to this the power of making herself practically invisible at will, and you will perceive that a Female, in Flatland, is a creature by no means to be trifled with.
>
> But here, perhaps, some of my younger Readers may ask *how* a woman in Flatland can make herself invisible. This ought, I think, to be apparent without any explanation. However, a few words will make it clear to the most unreflecting.
>
> Place a needle on the table. Then, with your eye on the level of the table, look at it side-ways, and you see the whole length of it; but look at it end-ways, and you see nothing but a point, it has become practically invisible. Just so is it with one of our Women. When her side is turned towards us, we see her as a straight line; when the end containing her eye or mouth—for with us these two organs are iden-tical—is the part that meets our eye, then we see nothing but a highly lustrous point; but when the back is presented to our view, then—being only sub-lustrous, and, indeed, almost as dim as an inanimate object—her hinder extremity serves her as a kind of Invisible Cap.
>
> The dangers to which we are exposed from our Women must now be manifest to the meanest capacity of Spaceland. If even the angle of a respectable Triangle in the middle class is not without its dangers; if to run against a Working Man involves a gash; if colli-sion with an Officer of the military class necessitates a serious wound; if a mere touch from the vertex of a Private Soldier brings with it danger of death;—what can it be to run against a woman, except absolute and immediate destruction? And when a Woman is invisible, or visible only as a dim sub-lustrous point, how difficult must it be, even for the most cautious, always to avoid collision!

What would a Flatland brain be like? Could it really work? Our 3-D brains are highly convoluted with microscopic neurons making complicated intercon-nections. Without this complex three-space network of nerve filaments—

something impossible to achieve on a plane without self-intersection—our brains would not be possible. However, as Martin Gardner notes in *The Unexpected Hanging*, one *can* imagine self-intersecting networks along which electrical impulses travel across intersections without turning corners.[7]

As Thomas Banchoff points out in *Beyond the Third Dimension*, the first edition of *Flatland* appeared as only one thousand copies in November 1884, but since then interest and sales have dramatically increased. Edwin Abbott Abbott was not the first person to consider a 2-D universe inhabited by flat creatures, but he was the first to explore what it would mean for 2-D creatures to interact with a higher-dimensional world. Today computer graphic projections of 4-D objects bring us a step closer to higher-dimensional phenomena, but even the most brilliant mathematicians are often unable to grasp the fourth dimension just as the square protagonist of Flatland had trouble understanding the third dimension.

Prior to Abbott's work, several individuals considered analogies between 2-D and 3-D worlds. For example, psychologist and physiologist Gustave Fechner wrote *Space Has Four Dimensions* in which a 2-D creature is a shadow man projected to a vertical screen by an opaque projector. The creature could interact with other shadows, but, based on its limited experience, could not conceive of a direction perpendicular to its screen. The idea of 2-D creatures dates back to Plato's *Allegory of the Cave*, in the seventh book of *The Republic*, where shadows are representations of objects viewed by 3-D observers constrained to watch the lower-dimensional views. Unlike Fechner, Plato does not suggest that the shadows have the capability of interacting with one another.

How to Hide from a Four-Dimensional Creature

Virtually all books about the fourth dimension suggest that it is impossible for us to hide from 4-D beings who could see inside our homes even with the doors closed. However, a lower-dimensional being could learn to hide from higher-dimensional beings by taking advantage of objects in the higher-dimensional world.

Let's start with a Flatland analogy. Consider a 3-D piece of Swiss cheese. If you were to drop this on Flatland so that it intersected the 2-D universe, a Flatlander could hide within the 2-D cross section of a hole in the cheese. The lucky Flatlander in this hole cannot be seen by a 3-D being. Of course, it might be difficult for the Flatlander to distinguish this hole from an ordinary 2-D circle; however, if there were some way for Flatlanders to sense and crawl into these holes, natural selection might evolve Flatlanders with such abilities. By analogy, we could hide from the prying eyes of a 4-D being by hiding in the hyperholes of a 4-D Swiss cheese or any 4-D object overlapping our world. Although we could not directly know the locations of 4-D hiding spaces, we might be able to infer their positions by noticing that there were spots where 4-D visitations never took place.

A Day at the Beach

Let's take a brief stroll along a 4-D beach filled with sunbathers. Our tour guide is a 4-D friend I'll call Mr. Plex. To take this tour, you need to be peeled out of our 3-D universe and placed into the fourth dimension. Look around. Much of what you see is confusing. Blobs appear out of nowhere, constantly changing in size, color, texture. Sometimes the blobs disappear and you can't tell which blobs are part of Mr. Plex and which are pieces of bodies from other bathers. Many of the blobs are flesh covered, so you let your imagination run wild, assuming that the bathing suits are quite scanty and you are watching a 4-D version of *Baywatch*.

Mr. Plex introduces you to his wife, Pamela Sue. You see a fleshy ball and another ball covered with blond hair. "Pleased to meet you," you say. Aside from the occasional hairy ball, the only way you can differentiate Mr. Plex from his wife is by observing how the blobs change shape. When Mr. Plex brings you to the snack bar, there is no way you can tell all the creatures apart. There are just too many changing blobs and colors.

Mr. Plex's artwork is strange and oddly disjointed both in space and in color combinations. You understand why. When you look at a 2-D painting on a wall, you step back in the third dimension and can see the boundary of the painting (usually rectangularly shaped) as well as every point in the painting. This means that you can see the entire painting from one viewpoint. If you wish to see a 3-D artwork from one viewpoint, you need to step back in the fourth dimension. Assuming that your eyes could grasp such a thing, you would theoretically see every point on the 3-D artwork, and *in* the 3-D artwork, without moving your viewpoint. This type of "omniscient" seeing and X-ray vision was known to Cubist painters such as Duchamp and Picasso. For this reason, Cubists sometimes showed multiple views of an object in the same painting. Present-day sculptors, such as Arthur Silverman, often place six copies of the same 3-D object, on separate bases, in six orientations. People viewing the six disjoint sculptures often do not realize that they are all the same object. Mathematics professor Nat Friedman (State University of New York at Albany) refers to this theoretical seeing in hyperspace as "hyperseeing" and points out in his writings that in hyperspace one can hypersee a 3-D object completely from one viewpoint. He calls a set of related sculptures with multiple orientations "hypersculpture." Friedman writes, "The experience of viewing a hypersculpture allows one to see multiple views from one viewpoint which therefore helps to develop a type of hyperseeing in our three-dimensional world." See Further Readings for a more complete description of hypersculpture given by Friedman.

Pinning God

In this chapter, we discussed the possibility of trapping a higher-dimensional being in our world by stabbing the creature's 3-D cross section using a knife. Interestingly, this idea of lower-dimensional confinement was the basis for "The Monster from Nowhere," a fascinating short story by Nelson Bond (see Further Readings). Burch Patterson, the hero of the adventure, travels to the Peruvian jungles searching for interesting animals. Suddenly, he and his men encounter a 4-D beast that appears to them as black blobs hovering in midair, disappearing and reappearing. Most of his men are attacked and killed. Others are lifted off the ground by the blobs and disappear into thin air. Patterson later determines that the pulsating blobs have dragged the men into a higher-dimensional universe.

Patterson yearns to capture the 4-D beast, but wonders how he can trap it. If he places a net around the beast, it can simply pull itself out of our universe and the net would fall to the ground empty. Patterson's strategy is to impale the blob with a spike so that it cannot leave our universe. This would be akin to a Flatlander's stabbing us and trapping us in the plane of its universe. After many weeks of studying the creature, Patterson identifies what he thinks is the beast's foot and drives a large, steel spike right through it. It takes him two years to ship the writhing, struggling blob back to New Jersey.

When the creature is exhibited to reporters, it struggles so hard that it tears its own flesh to escape and then proceeds to kill people and abduct Patterson into the fourth dimension. In the aftermath of the carnage, one of the survivors decides to burn all evidence of the beast so that no future attempts would be made to capture such a creature. It would simply be too dangerous.

> *Mulder:* Hey, Scully. Do you believe in the afterlife?
> *Scully:* I'd settle for a life in this one.
> —"Shadows," *The X-Files*

three

Surely it heard me cry out—for at that moment, like two exploding white stars, the hands flashed open and the figure dropped back into the earth, back to the kingdom, older than ours, that calls the dark its home.

—T. E. D. Kline, "Children of the Kingdom"

As a duly designated representative of the City, County and State of New York, I order you to cease any and all supernatural activities and return forthwith to your place of origin or to the nearest convenient parallel dimension.

—Ray Stantz in *Ghostbusters*

As one goes through it, one sees that the gate one went through was the self that went through it.

—R. D. Laing, *The Politics of Experience*

satan and perpendicular worlds

Washington Avenue, Washington, D.C., 6:00 P.M.

You and Sally are traveling in your car toward the White House. Snow continues to fall, and the asphalt roads start to break up. Soon sand begins to cover vast stretches of road.

You turn to Sally. "Sorry for the scare with the guinea pigs. I believe the Omegamorphs keep them as pets. The 4-D pigs seem to like the taste of breakfast cereals that I place in the jars every now and then."

Sally stares at you. "You mean they just materialize in your jars for treats. That's nuts."

You shake your head. "Haven't you ever wondered why breakfast cereals come in such big boxes and when you open brand-new boxes some of the cereal seems to have been removed? Sure, the cereal manufacturers tell you that the packages are sold by weight, not volume, and that the cereal settles over time. But the manufacturers are merely hiding the fact that the 4-D guinea pigs have entered the cereal boxes without breaking the cardboard. . . ."

"Watch your driving!" Sally yells as you begin to skid.

You wish you had snow tires. You shift into third gear, then second, as you travel more cautiously.

"Sally, could you turn on the radio?"

She presses the on button and your rear antenna automatically extends. As you scan for stations, all you hear are a lot of whispering sounds, almost as if there are sounds within sounds. Occasionally, there are a few quick screeches, but nothing you can quite understand.

"Sally, imagine you are about to be lifted up into the fourth dimension. What would our world look like to you from your higher perspective?"

"When we were in Cherbourg, I was apparently lifted up into hyperspace. I am loathe to accept such an outrageous explanation, but after seeing such odd things in your office, I'm beginning to think it's possible. Unfortunately, I fainted so I can't report anything."

"That's okay. We can explore the fourth dimension in the relative safety of my car." You hand Sally a large white card that you have pulled from your glove compartment. On it are impressive-sounding words written in capital letters:

AN *N*-DIMENSIONAL SPACE CUTS AN (*N* + 1) DIMENSIONAL SPACE INTO TWO SEPARATED SPACES

Sally flips the card over. "That's very erudite of you, but—"
"Yes?"
She stares at the card, wrinkled through years of use. "I'd think you'd impress more women with your FBI business card."
"Sally, do you know what the words on the card mean? Let me tell you."
You deliberately delay your answer as you give Sally a chance to admire your car seats made of Cordovan chamois leather—a luxurious, soft, porous leather that could be repeatedly wetted and dried without damage. Although your sporty, red Porsche Carrera XI is out of character with the spartan life of an FBI agent, you appreciate the car's sleek lines and blazing acceleration.

You would never have spent so much money on a car, but you've been able to obtain it for practically nothing. A few months ago while trolling in the Potomac River for murder victims, you hooked something big underwater. A day later you dived, saw the car, and had a friend tow it to shore. Mr. Duchovny, your boss at the FBI, said that because the vehicle identification numbers had been filed off, there was no way to trace the car. There was no sign of foul play—no blood stains or evidence of any kind except for a wet roll of hundred dollar bills you later found hidden beneath the spare tire. Because the police thought the car worthless after being under the river for a few months, they allowed you to keep it. Little did they anticipate your ingenuity for repair.

You look at Sally and press down the car's accelerator, hoping to hear Sally purr like a cat as the force of the engine pushes her ruthlessly back into the leather seat. Unfortunately, you do not get the desired effect.

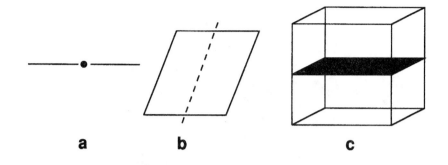

Figure 3.1 (a) A 0-D point cuts a 1-D line into two pieces. (b) A 1-D line cuts a 2-D plane into two pieces. (c) A 2-D plane cuts a 3-D space into two pieces. By analogy, our 3-D space would cut a 4-D hyperspace into two pieces.

She punches you in the arm. "You idiot. Keep your eyes on the road and drive carefully!"

You nod. "Okay, let me explain what I wrote on the card. Let's start in low dimensions and work up. Think of a point sitting on a 1-D line. Notice how the point cuts the line in two" (Fig. 3.1).

"Remember, we call the point zero-dimensional because there are zero degrees of freedom. If you lived in such a world, you could not move. Similarly, a 1-D line cuts a 2-D plane into two pieces."

Sally adjusts her seatbelt as you turn down Pennsylvania Avenue. "I can see where this is heading. A 2-D plane cuts a 3-D space into two pieces." On the card, she sketches an upper space and a lower space divided by a plane.

"Correct. And a 3-D space cuts a 4-D hyperspace into two pieces. In general, an n-dimensional space cuts an $(n + 1)$-dimensional space in half. For our discussion, I'll refer to the two regions of hyperspace, separated by the 3-D space, as located in the upsilon and delta directions. The words 'upsilon' and 'delta' can be used more or less like the words up and down. To cement the terms in your mind, think of Heaven as lying a mile in the upsilon direction, and Hell residing a mile in the delta direction. Of course, you can't see either living in our 3-D world."

You pull your car up to the northeast gate of the White House where you note the flashing lights of police cars. Despite your curiosity, you continue your lecture to make sure Sally has a firmer grasp on the fourth dimension. "Some religions suggest that Satan is the devil, the fallen

Figure 3.2   Lucifer falling from Heaven (one mile in the upsilon direction) down into Hell (one mile in the delta direction). (Drawing by Brian Mansfield.)

prince of angels, the adversary of God. Satan was said to have been created by God and placed at the head of the angelic hosts, where he enticed some of the angels to revolt against God. In punishment for his rebellion, Satan was cast from Heaven together with his mutinous entourage, which were transformed into demons. Today, if Satan were to fall from upsilon to delta through our 3-D space, what might we see?"

Sally gazes out the window toward the White House and then back at you. "It would be similar to what a Flatlander what might see if a man was cast down from above to below the plane of Flatland. We might see a shocking movement of horrible, incomprehensible cross sections of Satan, moving, merging, and spreading. If a 3-D Satan had horns and a tail and fell through Flatland, these body parts might be perceived as fleshy or bony circles changing in size."

"Right. Satan's fall would appear to us as incomprehensible fleshy blobs suddenly materializing, changing in size, and finally disappearing" (Fig. 3.2).

Sally adjusts the little golden crucifix she wears on a thin chain around her neck. "When I went to Catholic school, the nuns taught me that

Satan is the ruler over the fallen angels, always struggling against the Kingdom of God by seducing humans to sin. Satan disrupts God's plan for salvation and slanders the saints to reduce the number of those chosen for the Kingdom of God. How might Satan do this from delta below our 3-D world? If he could occasionally pop up into our world, he would wreak havoc!"

"Of course, we're only talking about metaphors here. But this would explain our encounter at Cherbourg. We saw pulsating bags of skin when a 4-D creature came into our world."

Sally leans forward. "Was that Satan at Cherbourg?"

"That's what we're here to find out. I think it was an Omegamorph. Lots of mysterious things are going on these days in Cherbourg and around Washington, D.C. Nothing is certain. I already mentioned that if we fell through Flatland, the Flatlanders would see pulsating, pancake-like blobs of skin as we intersected their world. When our mouths intersected Flatland, Flatlanders might get a glimpse of the edge of our tongues, which would be like bumpy pink shapes to them. When our heads fell through, they'd see the edges of a hairy pancake" (Fig. 3.3).

"What would it look like to be lifted upsilon into the fourth dimension?"

"Sally, think of a 3-D creature pulling on a 2-D square. If the square were slowly peeled from the 2-D world, part of the square would remain in the plane for a time. Similarly, if a 4-D creature lifted you from the 3-D world, your head might disappear while your body remained. And then you would be entirely lifted from the world."

"One thing is bothering me."

"Just one?"

"I understand that a 4-D being should be able to glimpse into our guts, see the valves of our heart, and so forth, but what might their eyes be like? How would their eyes function?"

"Sally, our own retina is essentially a 2-D disc of rods and cones—the two kinds of nerve cells for vision. By analogy, a 4-D creature would have a retina that was a sphere of nerve endings."

"How would an Omegamorph see with a spherical retina?"

"When you look at a circle moving in Flatland, your 2-D retina captures a circular pattern impinging on the nerve cells. Each point in the circle corresponds to a light ray from the circle to a single point in your retina. If you are looking from above, each ray goes up to your eye. If a 4-

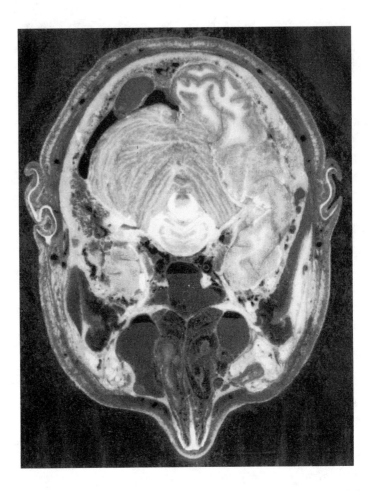

Figure 3.3 A Flatlander would see the edge of your head as it intersected their universe. (In this slice of a fresh cadaver, we can see the cerebellum, cerebral cortex, brainstem, and nasal passages, although a Flatlander would just see the outer edge of the cross section.)

D being were looking at a ball, the resultant image is produced by the excitation of a ball-shaped pattern of nerve endings in the spherical retina. Each point on the 3-D ball sends a light ray upsilon to a ball-shaped region in the spherical retina" (Fig. 3.4).

Sally stares out the window and then back at you. "Here's where my medical knowledge comes in handy. To help us see our world, we have an army of cells in our visual system to enhance contrast, detect motion and

Figure 3.4   The retina of a 4-D being's eye.

edges, and distinguish between small gradations in intensity and color. In fact, there are many millions of cells to aid us in gaining insight about our world: 125 million rods and cones, several million intervening cells in the retina, and one million neurons in the lateral geniculate nucleus, the first major processing station for visual input. If we have 125 million rods and cones in our 2-D retina, I'd expect roughly 1.3 trillion cells in the 3-D retina of a 4-D being's eye."

"Wow, how did you come up with that?"

"You need to first take the square root of 125 million to estimate the number of cells in a single dimension, and then cube that. Because the retinas are one dimension less than the dimension of the creature, we could estimate the number of cells in all higher-dimensional retinas."

You glance out the window and sense something moving nearby. You lift your head but see nothing unusual. You roll down the window, but that doesn't help. Should you step out of the car to look?

And then he appears. An old man in a Santa Claus outfit. He seems to tower above the crowds, but strangely none of the police find him peculiar. Yet the man is clearly out of place. For some reason, you feel embarrassed that the man is so obviously seeking attention. You wish the man would go away. He does not belong. It's a feeling you frequently have.

You roll up the window. "Sally, I think I can help you understand how a 3-D retina can visualize a human, inside and out, at the same time. First consider a blood transfusion for 2-D creatures via a tube that goes up into the third dimension and back down again into the plane of the creature. As far as the creature is concerned, you are not breaking the skin, which might be represented as a line in a drawing. Can you visualize how 4-D creatures can transfuse us?" (Fig. 3.5).

"Yes, they could transfuse a 3-D creature by using a transfusion tube that goes upsilon and delta into the fourth dimension without ever breaking the skin. The same concept applies to stealing objects in safes without ever breaking the safe wall. Your hand would materialize in the safe and then dematerialize as you withdrew it. But where does the transfusion tube really go?" (Fig. 3.6).

"It goes upsilon where the operating room doesn't even exist! Similarly, a 4-D creature's 3-D retina can see all of your insides without breaking your skin. This assumes that light rays are reflected into the fourth dimension—we'll have to research that further. I've memorized Abbott's nice description of a 2-D creature lifted out of Flatland and looking down into his 2-D world from a 3-D world":

I felt myself rising through space. It was even as the Sphere had said. The further we receded from the object we beheld, the larger became the field of vision. My native city, with the interior of every house and every creature therein, lay open to my view in miniature. We mounted higher, and lo, the secrets of the earth, the depths of mines and inmost caverns of the hills, were bared before me.

Sally nods. "We've talked about how 4-D beings would look to us, but what would we *really* look like to them?

"To answer that question, let's talk about the appearance of creatures living in worlds perpendicular to one another."

"Perpendicular worlds?"

"Yes, again I'll start with a 2-D analogy. Consider two planes on which thousands of intelligent insects live. Their two worlds intersect in a line. Along the line of intersection are many moving line segments changing size, disappearing and reappearing" (Fig. 3.7).

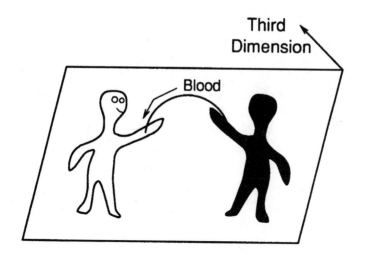

Figure 3.5 A "hyperspace" blood transfusion in a 2-D world. The transfusion tube goes up and down into the third dimension. The creature's skin is never broken.

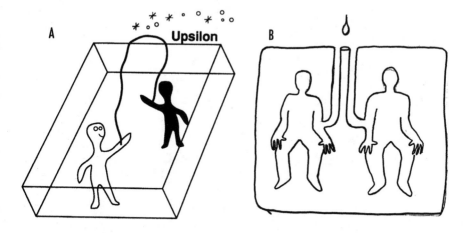

Figure 3.6 A hyperspace blood transfusion in a 3-D world. (a) The transfusion tube goes upsilon and delta into the fourth dimension. (b) Artist's interpretation of a hyperspace transfusion. (Drawing by Clay Fried.)

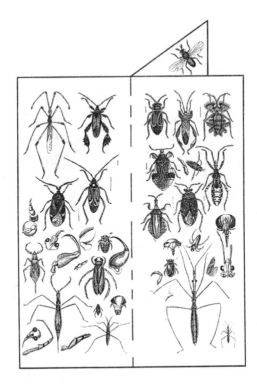

Figure 3.7 Perpendicular insect worlds.

You turn up the Porsche's heater. Perhaps both of you should get out of the car and find out what's happening in the White House. "Sally, in four dimensions, it is possible to have two 3-D spaces perpendicular to each other. They would have a plane in common."

"If there were a 3-D space perpendicular to ours, a space with aliens moving around in it, how would they appear to us? Could we even see them at all?"

"I am in touch with this other world and will now reveal to you the answer."

You pull down the sun visor above your seat, revealing a photo of actress Gillian Anderson.

"What's she doing here?"

"Don't worry. The photo simply conceals a button that will let you glimpse a plane intersection with a 3-D world perpendicular to ours." You press on the photo and suddenly a rectangular plane rises up from the car's dashboard. "This is the intersection of a 3-D space of aliens with our 3-D space."

Drifting around the plane are various thin blob shapes.

Sally reaches out to touch one. "This is amazing! They feel solid" (Fig. 3.8).

"You're correct."

Some bigger fleshy blobs suddenly appear and drift on the plane.

"What are those?"

"Perhaps a big creature just walked by."

"Okay, but I'm still wondering what we would see if we were lifted upsilon into hyperspace and could gaze at our world. I want to know what it feels like to be a God."

"Sally, let's consider a 2-D analogy with a man living on Flatland. If we were to rip him off his plane of existence and up into our 3-D space, I don't think he would really be able to see all the 2-D objects in his world as we do—assuming his retina stays the same and is only a 1-D arc, a line segment in the back of his eye. His retina is designed and evolved to receive images in the plane of Flatland. If he were to look down on the plane of Flatland, it would be as if he were scanning the plane through a thin slit, so he wouldn't really be able to see his world all at once. As he moved his head back and forth, like a supermarket barcode scanner, different regions of his world would come into view. Perhaps he could even put all the 1-D images together to visualize his planar world—if his brain were sufficiently versatile. The situation is similar to the insects on the two perpendicular planes. As they gaze along their plane of sight, they would see moving line segments. If one plane moved relative to the other, they would be scanning different regions of a plane. Likewise, when you were ripped out and upsilon into hyperspace, and gazed delta onto our world, you could have seen (if you did not faint!) many planar cross sections of our world. You'd see the insides and outsides of things. If you tried hard enough, you might be able to combine all these sections into a composite 3-D image of everything."

"If God existed in the fourth dimension and gazed delta into our world, would He only be seeing planar cross sections? That sounds limiting."

"No, remember our retina is two-dimensional. His retina could be three-dimensional, allowing Him to see everything at once."

Sally runs her fingers lingeringly on the soft leather of her seat. "Maybe we could learn to 'see' 3-D objects if we were gazing delta on to our world. But how can we ever visualize 4-D objects?"

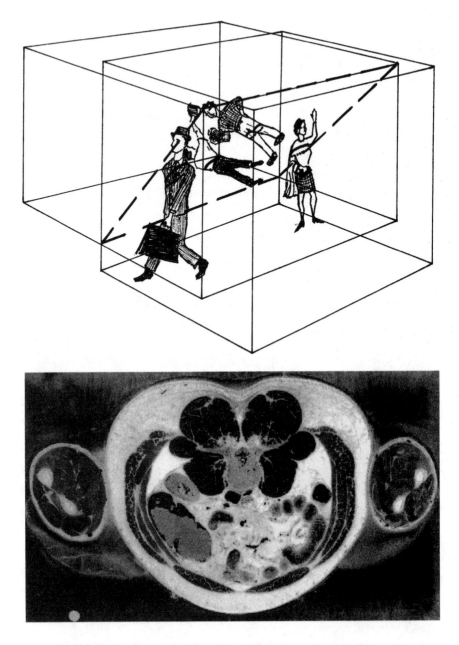

Figure 3.8 3-D alien world, perpendicular to ours and intersecting our world at a plane. (top) Visualization of our world and theirs. (bottom) A human in their world appears as a cross section when viewed from our world.

"Professor Rudy Rucker of San Jose State University writes philosophically on this precise topic":

Drawings use 2-D arrangements of lines to represent 3-D objects. Why shouldn't we be able to build up 3-D arrangements of neurons that represent 4-D objects? More fancifully, perhaps our minds are not just 3-D patterns: maybe our brains have a slight 4-D hyperthickness; or maybe our minds extend out of our brains and into hyperspace!

You reach into your glove compartment and remove a gray, wrinkled thing stored in a formalin-filled jar to prevent decay. You give a little tap on the jar marked "Einstein." His cerebrum jiggles like a nervous mango. "Sally, the mammalian brain seeks to expand its dimension to fulfill its biological purpose. For example, the huge 2-D surface of the brain is intricately folded to fill a 3-D volume in order to increase its surface area. Wouldn't it be a great science-fiction story that describes a human whose 3-D brain folds itself in the fourth dimension to increase its capacity?" (Fig. 3.9).

You place the brain back into the glove compartment and withdraw a chunk of Swiss cheese filled with tunnels that interconnect various regions of the cheese.

Sally pinches her nose. "No thanks."

"Sally, maybe we could use other spatial dimensions for wormhole travel."

"Wormholes?"

You nod. "Someday, wormholes might be used for wonderful journeys. Some physicists believe that at the heart of all space, at submicroscopic size scales, there exists *quantum foam*. If we sufficiently magnified space, it would becomes a seething, probabilistic froth—a cosmological cheese of sorts."

Sally runs her fingers through her hair. "Now, this sounds quite interesting. What do we know about the quantum froth?"

Your heartbeat increases in frequency and amplitude as you gaze at her skirt, the color of cool mint. "In the froth, space doesn't have a definite structure. It has various probabilities for different shapes and curvatures. It might have a 60 percent chance of being in one shape, a 20 percent chance of being in another, and a 20 percent chance of being in a third form. Because any structure is possible inside the froth, we can

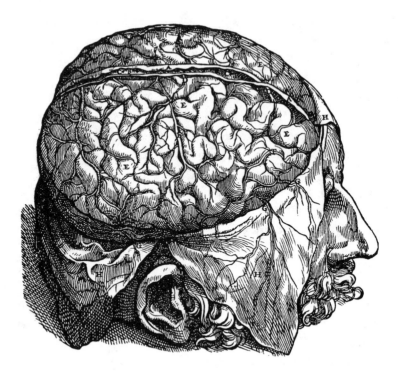

Figure 3.9   The mammalian brain seeks to expand its dimension to fulfill its biological purpose. For example, the huge 2-D surface of the brain is intricately folded to fill a 3-D volume so that its surface area will increase. If you laid all your brain cells end to end, they'd stretch around the world twenty-five times.

call it a *probabilistic foam*, or *quantum foam*. The froth contains holes to other regions in space and time."

Sally taps on the dashboard. "Could these wormholes be connecting different areas of our same universe?"

"Yes. In the foam, adjacent regions of space are continually stealing and giving back energy from one to another. These cause fluctuations in the curvature of space, creating microscopic wormholes. Who knows, someday civilizations might be able to use such wormholes to travel the universe."

The tapping of Sally's fingers becomes more incessant. "But those wormholes are too tiny for people to use."

"That's the problem, and why we might never be able to use a wormhole. We'd need a device that spews out something called *exotic matter*.

Exotic matter has special properties that will enlarge and hold open a wormhole. Maybe some advanced extraterrestrial civilization has such a device, but we don't know."

You motion to all the police cars with flashing lights. "Let's find out what's happening here." You get out of the car and walk over the manicured grounds toward the White House entrance.

A large man wearing a Secret Service–style radio earpiece and dressed in a black suit with a hundred-dollar haircut is standing dead center in front of the entrance with his finger pointing at you. The man is at least six feet five, 250 pounds, with a neck twice the thickness of yours, and a nose that has been broken more times than you care to imagine. He seems one part football player and one part weight lifter. "No one enters the White House," he says.

You flash your FBI badge.

The man shakes his head. "No one enters the White House."

You sigh. This conversation wasn't going to be productive. "Whose house is this?" you ask.

The man looks confused by that one.

You point to yourself. "My good man, the answer to that question is simple—the house is mine. I'm a taxpayer. I'm also from the FBI. Moreover, I'm here to protect the president from the Omegamorphs." You feign left, cut right, and step into the White House with Sally close behind.

The man puts his huge ham hand on your shoulder. This is not a wise choice—even for someone so much larger than you. You have the odd but compelling idea that your FBI badge entitles you to investigate crimes.

You countergrab the man's hand. The palm of your right hand lifts the man's arm at his elbow joint, causing the arm to hyperextend, and then you form a chicken-beak shape with your right hand as you swipe the man's arm away.

Suddenly there is a scream from within the White House. You race through the East Room and then through the Red Room. You dash past paintings of Abraham Lincoln, Zachary Taylor, and John F. Kennedy. Finally, arriving at the State Dining Room, you find the president. He is surrounded by blobs, obviously one or more creatures from the fourth dimension.

The president's Secret Service surround him, pointing their weapons at the fleshy blobs, but it's hard to get a clean shot at shapes constantly changing size and disappearing and reappearing.

One of the Secret Service agents looks at you. "Who the hell are you?"

You reach into your pocket and withdraw your badge. "FBI."

The president screams as his hand disappears.

"Get some rope," you scream to the Secret Service agents. "We've got to anchor him in our world. Right now an Omegamorph is trying to pull him into the fourth dimension."

The nearest Secret Service agent looks at you. "Are you some kind of nut?"

Sally withdraws her FBI badge and points her service revolver at the undulating bags of flesh. "Listen to him. He knows what he's talking about."

One of the agents dashes into a storage closet and brings out a thick rope. "Is this good enough?"

You grab the rope, rush toward the president, and quickly loop the rope around his ankle. "Sally, quick, tie the other end to the bust."

She nods and swiftly attaches the rope's other end to a huge bronze bust of Abraham Lincoln.

The president's arm disappears. "What do they want with me?"

You try to hold on to the president. "Sir, we're in a region of space where it's easy to peel people from the local 3-D space."

The president screams as his head disappears, then his chest and legs. The only part of his body you see is his foot, tied to the rope, and floating several inches off the ground. His disembodied foot dances and finally slips out of the knot.

"No!" Sally yells.

You slowly shake your head as a tear forms in your eye.

The president of the United States has been abducted upsilon into hyperspace.

The Science Behind the Science Fiction

The soul is trapped within the *cage* of the body. The awakened soul can progress along a way which leads to annihilation in God.

—Afkham Darbandi and Dick Davis,
Introduction to *The Conference of the Birds*

There are some qualities—some incorporate things,
That have a double life, which thus is made
A type of twin entity which springs
From matter and light, evinced in *solid* and *shade.*
—Edgar Allan Poe, "Silence"

If 4-D analogs of unclothed humans intersected our world, they might appear as fleshy or hairy blobs (Fig. 3.10), as described in this chapter. If clothed, we would see cloth-covered bags of flesh. (Artist Michelle Sullivan has drawn several fanciful illustrations of 4-D beings in Fig. 3.11.) By analogy, if you were to stick your foot into Flatland, a Flatlander would see a leather disc (a cross section of your shoe). Without the shoe, the Flatlander would see a fleshy disc (your skin). As you stuck your foot deeper into Flatland, the Flatlander would see a fabric disc representing your pants. If you stuck both legs into Flatland, you would appear as two fabric discs. As you drifted downward, these two discs merge into one disc at your waist and then change colors and break apart into three discs (your two arms and shirt). As you descended, the Flatlander would finally see a hairy disc (the hair on your head) that suddenly disappears as you go all the way through Flatland. To Flatlanders, you would be something out of their worst nightmares—a confusing collection of constantly changing discs made of leather, cloth, flesh, lips, teeth (when your mouth is open), and hair.

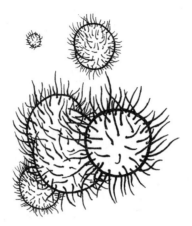

Figure 3.10 Artistic rendition of a 4-D being who appears in our world as a set of hairy flesh-balls. (Drawing by Michelle Sullivan.)

Figure 3.11 Fanciful depictions of 4-D beings by artist Michelle Sullivan. Michelle often designs her creatures with disjoint parts to symbolize the intersection of a 4-D creature in our 3-D world.

Similarly, if a 4-D being came into our world, we might see a maddening collection of constantly changing balls made of leather, cloth, flesh, hair, and dental enamel. Just imagine waking up one day only to find a white enamel ridged ball in your bedroom. It is the intersection of an Omegamorph's tooth in the 3-D world. This 4-D being thinks you are a dentist and wants you to drill and fill his cavity. Could you successfully drill out the cavity from your location in the 3-D universe?

Parallel Universes

Although I deal with the fourth spatial dimension in this book, many science and science-fiction authors have used the word "dimension" when referring to parallel universes—universes that resemble our own and perhaps even occupy the same space as our own. Small differences in the evolution of such worlds can lead to strangely different universes as the universes evolve. For example, imagine a slight variant of our world in which Cleopatra had an ugly but benign skin growth on the tip of her nose. The entire cascade of historical events would be different. A mutation of a single skin cell caused by the random exposure to sunlight will change the universe. This entire line of thinking reminds me of a quote from writer Jane Roberts:

> You are so part of the world that your slightest action contributes to its reality. Your breath changes the atmosphere. Your encounters with others alter the fabrics of their lives, and the lives of those who come in contact with them.

In her novel *Memnoch the Devil*, Anne Rice has a similar view when she describes heaven:

> The tribe spread out to intersperse amongst countless families, and families joined to form nations, and the entire congregation was in fact a palpable and visible and interconnected configuration! Everyone impinged upon everyone else. Everyone drew, in his or her separateness, upon the separateness of everyone else!

In some science-fiction scenarios where you can freely travel between parallel universes, it's easy to create duplicates of yourself. For example, consider several universes that are identical until you travel from one universe to another. You leave Universe 1, travel to Universe 2, and live for a year in Universe 2 with your replica in Universe 2. Assume there is a Universe 3 virtually identical to Universe 2 that has two copies of you. You leave Universe 2 and travel to Universe 3, and so on. (One of my favorite tales of replication is David Gerrold's *The Man Who Folded Himself.*) By repeating such loops, you can create as many replicas as you like. The universe gets complicated, but there are no logical contradictions. Although this multiple-universe concept may seem far-fetched, serious physicists have considered such a possibility. In fact, Hugh Everett III's doctoral thesis "Relative State Formulation of Quantum Mechanics" (reprinted in *Reviews of Modern Physics*) outlines a controversial theory in which the universe at every instant branches into countless parallel worlds. However, human consciousness works in such a way that it is only aware of one universe at a time. This is called the "many-worlds" interpretation of quantum mechanics. One version of the theory holds that whenever the universe ("world") is confronted by a choice of paths at the quantum level, it actually follows both possibilities, splitting into two universes. These universes are often described as "parallel worlds," although, mathematically speaking, they are orthogonal or at right angles to each other. In the many-worlds theory, there may be an infinite number of universes and, if true, then all kinds of strange worlds exist. In fact, some believe the controversial notion that somewhere virtually everything must be true. Could there be a universe where fairy tales are true, a real Dorothy lives in Kansas dreaming about the Wizard of Oz, a real Adam and Eve reside in a Garden of Eden, and alien abduction really does occur all the time? The theory could imply the existences of infinite universes so strange we could not describe them. My favorite tales of parallel worlds are those of Robert Heinlein. For example, in his science-fiction novel *The Number of the Beast* there is a parallel world almost identical to ours in every respect except that the letter "J" does not appear in the English language. Luckily, the protagonists in the book have built a device that lets them perform controlled explorations of parallel worlds from the safety of their high-tech car. In contrast, the protagonist in Heinlein's novel *Job* shifts through parallel worlds without control. Unfortunately, just as he makes some money in one America, he shifts to a slightly different America where his money is no longer valid currency, which tends to make his life miserable.

The many-worlds theory suggests that a being existing outside of spacetime might see all conceivable forks, all possible spacetimes and universes, as always having existed. How could a being deal with such knowledge and not become insane? A God would see all Earths: those where no inhabitants believe in God, those where all inhabitants believe in God, and everything in between. According to the many-worlds theory, there could be universes where Jesus was the son of God, universes where Jesus was the son of the devil, and universes where Jesus did not exist. (See Addendum.)

Much of Everett's many-worlds interpretation is concerned with events on the submicroscopic level. For example, the theory predicts that every time an electron either moves or fails to move to a new energy level, a new universe is created. Currently, it is not clear the degree to which quantum (submicroscopic) theories impact on reality at the macroscopic, human level. Quantum theory even clashes with relativity theory, which forbids faster-than-light (FTL) transfer of information. For example, quantum theory introduces an element of uncertainty into our understanding of the universe and states that any two particles that have once been in contact continue to influence each other, no matter how far apart they move, until one of them interacts or is observed. In a strange way, this suggests that the entire universe is multiply connected by FTL signals. Physicists call this type of interaction "cosmic glue." The holy grail of physics is the reconciling of quantum and relativistic physics.

What exactly is quantum theory? First, it is a modern science of the very small. It accurately describes the behavior of elementary particles, atoms, molecules, atom-sized black holes, and probably the birth of the universe when the universe was smaller than a proton. For more than a half-century, physicists have used quantum theory as a mathematical tool for describing the behavior of matter (electrons, protons, neutrons . . .) and various fields (gravity, weak and strong nuclear forces, and electromagnetism). It's a practical theory used to understand the behavior of devices ranging from lasers to computer chips. Quantum theory describes the world as a collection of possibilities until a measurement makes one of these possibilities real. Quantum particles seem to be able to influence one another via quantum connections—superluminal links persisting between any two particles once they have interacted. When these ultrafast connections were first proposed, physicists dismissed them as mere theoretical artifacts, existing only in mathematical formalisms, not in the real world. Albert Einstein considered the idea to be so crazy that it had to demonstrate there was something missing in quantum theory. In the late 1960s, however, Irish physicist John Stewart Bell proved that a quantum connection was more than an interesting mathematical theory. In particular, he showed that real superluminal links between quantum particles explain certain experimental results. Bell's theorem suggests that after two particles interact and move apart outside the range of interaction, the particles continue to influence each other instantly via a real connection that joins them together with undiminished strength no matter how far apart the particles travel. Alain Aspect and his colleagues confirmed that the property is in fact an actual property of the real world. However, the precise nature of this faster-than-light quantum connection is still widely disputed.

Wormholes

Much of the recent research on wormholes has been conducted by Kip Thorne, a cosmologist, and Michael Morris, his graduate student. In a scientific paper published in the *American Journal of Physics* (see Further Readings), they developed a theoretical scheme for inter- and intra-universal travel via wormholes bridging the universe. These cosmic gateways might be created between regions of the universe trillions of miles apart and would allow nearly instantaneous communication between these regions.

Appendix B lists numerous uses of the fourth dimension, hyperspace, and wormholes in science fiction. For example, Carl Sagan in his novel *Contact* also uses the Kip Thorne wormholes to traverse the universe. The television shows *Star Trek: The Next Generation*, *Star Trek: Voyager*, and *Star Trek: Deep Space Nine* have all used wormholes to travel between faraway regions of space. In *Star Trek: Deep Space Nine*, a station stands guard over one end of a stable wormhole.

Cosmic wormholes created from subatomic quantum foam were also discussed by Kip Thorne and his colleagues in 1988. Not only did these researchers claim that time travel is possible in their prestigious *Physical Review Letters* article, but time travel is *probable* under certain conditions. In their paper, they describe a wormhole connecting two regions that exist in different time periods. Thus, the wormhole may connect the past to the present. Because travel through the wormhole is nearly instantaneous, one could use the wormhole for backward time travel. Unlike the time machine in H. G. Wells's *The Time Machine*, the Thorne machine requires vast amounts of energy—energy that our civilization cannot possibly produce for many years to come. Nevertheless, Thorne optimistically writes in his paper: "From a single wormhole an arbitrarily advanced civilization can construct a machine for backward time travel."

Note that the term "wormhole" is used in two different senses in the physics literature. The first kind of wormhole is made of quantum foam. Because of the foam-like structure of space, countless wormholes may connect different parts of space, like little tubes. In fact, the theory of "superspace" suggests that tiny quantum wormholes must connect every part of space to every other part! The other use of the word "wormhole" refers to a possible zone of transition at the center of a rotating black hole.

To fully realize a wormhole, Morris and Thorne calculated various properties of matter required to form the wormhole's throat. One property of interest to them was the tension (i.e., the breaking strength) of matter needed to keep the wormhole open. What they found was that the required tension would be very large. As Paul Halpern in his book *Cosmic Wormholes* notes, for a throat that is four miles across, the quantity of force needed is 10^{33} pounds per square inch. This would be more than the pressure of a trillion boxes, weighing a trillion tons each, placed in the palm of your hand. Larger wormholes with wider throats would have more reasonable values for throat tension.

In additional, Morris and Thorne found another difficult situation to overcome when considering the matter needed to form the gateway. The tension required for keeping the wormhole open must be 10^{17} times greater than the density of the substance used to build the wormhole. According to current science, there is no matter in the universe today having breaking tensions so much larger than their densities. In fact, if the tension of a piece of matter were to rise above 10^{17} times its own density, physicists feel that the material would begin to possess strange attributes, such as negative mass. Because of these unusual characteristics, the type of matter needed to keep wormholes open has been called *exotic matter*. Matter of this type may exist in the vacuum fluctuations of free space. To make wormhole construction easier, it may be possible to construct the entire wormhole out of normal matter and use exotic matter only in a limited band at the throat.[1]

Hyperdimensional Chess Knights and Monopoly

Let's have some real fun in these last sections. Not only is it interesting to speculate about the fourth dimension in mathematics and physics, but the fourth dimension also provides a fertile ground for extending puzzles and games. As an example, let's consider chess.

Chess is essentially a 2-D game in which pieces slide along the surface of the checkerboard plane. Playing pieces usually can't jump up into the third dimension to get around one another. The Knight, however, is a hyperdimensional being because it can leave the playing board plane to leap over other pieces in its way. (The Knight is "hyperdimensional" in the sense that it can exploit the

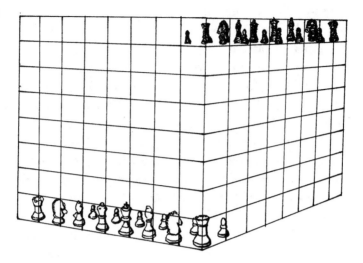

Figure 3.12 Extending chess games into higher dimensions. (Drawing by Brian Mansfield.)

third dimension whereas most other pieces are constrained to a 2-D world.) Are there other hyperdimensional chess pieces? I once invented a truly 2-D game where Knights could not leave the plane to get over other pieces. Try it. How does this affect your strategy? Could a world chess master constrained to using 2-D Knights defeat an excellent player using standard hyperdimensional Knights? What would have been the outcome for the game played between chess master Gary Kasparov and IBM's Deep Blue computer if Kasparov had a hyperdimensional Rook that could jump over other pieces?

It is also possible to extend chess to higher dimensions by substituting higher-dimensional boards for the standard 8 × 8 square board. For example, chess can be generalized to three dimensions by playing on a 8 × 8 × 8 grid of positions in a 3-D cube. (You can build a physical model[2] of a 3-D array of cubes, with each a possible position for a chess piece, or use computer graphics to create a virtual playing board.) Can you design such a board and modify some of the chess moves so that they extend into the third dimension (Fig. 3.12)? For example, the Queen might also move diagonally in the third dimension. Is it difficult to checkmate a King that might move away to twenty-six positions? What discoveries can you make about game strategies and the relative power of chess pieces? Generalize your results to the fourth dimension. Also consider Möbius chess played on a Möbius band (Fig. 3.13). (As you will

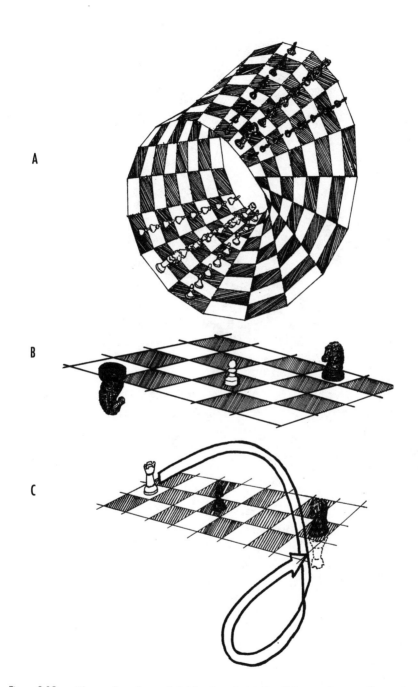

Figure 3.13 Chess played on a Möbius band. (a) Possible starting configuration. (b) In Möbius chess, either Knight can attack the pawn. (c) In this configuration, the pawn does not necessarily protect the Knight, because the Rook may travel in the opposite direction and end up beneath the Knight. (Drawing by Brian Mansfield.)

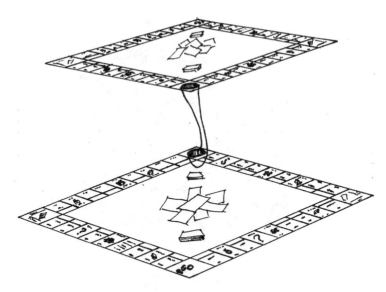

Figure 3.14 Extending Monopoly games into higher dimensions. The Free Parking square acts as a wormhole between parallel universes. (Drawing by Brian Mansfield.)

learn in Chapter 5, a Möbius strip is created by twisting a strip of paper 180 degrees and then taping the ends together. The resultant object has only one side.) Now that chessmaster Gary Kasparov has been beaten by an IBM computer, perhaps chess variants such as these would provide an infinite reservoir for new matches and theories.[3]

When I was young, I played hyperdimensional Monopoly games by aligning two or more boards, side by side (Fig. 3.14). The Free Parking square acted as a "wormhole" between parallel universes. In other words, if a playing piece lands on Free Parking, it has the option of moving to an adjacent board. Try this mind-boggling variant. I'd be interested in hearing from readers who have experimented with hyperdimensional Monopoly.

Rubik's Tesseract

Many of you will be familiar with Ernö Rubik's ingenious cubical puzzle and its variations that include a 4 × 4 × 4 cube and puzzles shaped like tetrahedra. One natural variation that never appeared on toy store shelves is the 4-D version of Rubik's cube—Rubik's tesseract. Dan Velleman (Amherst College) discusses the 3 × 3 × 3 × 3 Rubik's tesseract in the February 1992 issue of *Mathematics Magazine*. Many of his findings were discovered with the aid of a colorful simulation on a Macintosh computer. Velleman remarks, "Of course, the tesseract is somewhat harder to work with than the cube, since we can't build a physical model and experiment with it." Those of you interested in pursuing the details of this mind-shattering tesseract should consult his paper. See Appendix A for more information on Rubik's tesseract.

> *Scully:* I forgot what it was like to spend a day in court.
> *Mulder:* That's one of the luxuries of hunting down aliens and genetic mutants. You rarely get to press charges.
> —"Ghost in the Machine," *The X-Files*

four

The sphere of my vision now began to widen. Next I could distinctly perceive the walls of the house. At first they seemed very dark and opaque, but soon became brighter, and then transparent: and presently I could see the walls of the adjoining dwelling. These also immediately became light, and vanished— melting like clouds before my advancing vision. I could now see the objects, the furniture, and persons, in the adjoining house as easily as those in the room where I was situated. . . . But my perception still flowed on! The broad surface of the earth, for many hundred miles, before the sweep of my vision—describing nearly a semicircle—became transparent as the purest water; and I saw the brains, the viscera, and the complete anatomy of animals that were at the moment sleeping or prowling about in the forests of the Eastern Hemisphere, hundreds and even thousands of miles from the room in which I was making these observations.

—Andrew Jackson Davis, *The Magic Staff*

Consider the true picture. Think of myriads of tiny bubbles, very sparsely scattered, rising through a vast black sea. We rule some of the bubbles. Of the waters we know nothing.

—Larry Niven and Jerry Pournelle, *The Mote in God's Eye*

hyperspheres and tesseracts

"We've got to save the president!"

"Sally, we'll try. But first we have to continue our lessons. We need more insight before going after him."

"We don't have time for that."

"Sally, we must make the time. Christopher Columbus didn't start exploring without first understanding basic principles of navigation."

You've returned to your FBI office. "I want to talk more about hyperspheres and tesseracts, the 4-D counterparts to spheres and cubes." You draw a circle on the board with a dot at its center. "A circle is the collection of points (on a plane) all at the same distance r from a point. A sphere is the collection of points (in space) all at the same distance r from a point. Similarly, a hypersphere is the collection of points (in hyperspace) all at the same distance r from a point."

Sally steps closer to the board. "If a hypersphere with a seventeen-foot radius came into our world, what points would be on it?" She points her finger in the air. "Assume that its center is located at my fingertip."

For several seconds, you stare at Sally's elegantly manicured nails before taking a string from your drawer and measure off seventeen feet. "Sally, may I borrow one of your earrings?"

She removes a small, golden earring from her left ear and hands it to you. You tie a knot around it. You tie the other end of the string to her finger and walk away until the string is taut. "All the points that are seventeen feet from your fingertip would be on the hypersphere. Right now, your earring would touch the hypersphere's edge. As long as the string is taut, the ring stays on the hypersphere as I move the ring through our

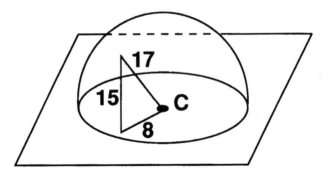

Figure 4.1 A cross section of a hypersphere centered at C. The radius of the hypersphere is seventeen feet.

space. But suppose we could move the string upsilon into the fourth dimension. I could move the earring eight feet away from your finger in our space, then turn at right angles, and then move fifteen feet upsilon into hyperspace. Your ring would still be on the hypersphere."

"How did you determine that eight and fifteen were the correct numbers?"

"Do you remember the Pythagorean theorem, or distance formula? If you move only in two directions, the distance d would be measured by $x^2 + y^2 = d^2$. So for our example, $8^2 + 15^2 = 17^2$" (Fig. 4.1).

Sally nods. "Using your formula, this means that no matter in what direction you move eight feet away from my finger, the additional fifteen-foot move upsilon gives a point exactly seventeen feet away from my fingertip."

"Yes. That also means if we take all the points on an eight foot sphere around your fingertip, and them move upsilon fifteen feet, we will get a displaced sphere of points all belonging to the seventeen-foot hypersphere around your fingertip."

Sally thinks for a few seconds. "Now I can understand why a hypersphere consists of a series of spheres—spheres that grow smaller as one moves upsilon or delta away from my fingertip at the sphere's center. Also, the less I move away from the hypersphere's center in our space, the more I can move upsilon or delta to be on the hypersphere's surface."

"That's right. All the spheres make up a 3-D hypersurface that's analogous to a 2-D surface of a sphere. The hypersurface of a hypersphere is simply 3-D space curved in 4-D space. As I once told you, some scientists think that our universe is the hypersurface of a very large hypersphere. However, even if this is true, people seem to be confined to three degrees of freedom because they are confined to the 3-D surface. Similarly, an ant walking on the surface of a sphere is confined to the 2-D surface and has two degrees of freedom.

You go over to a handsome walnut cabinet in a dark corner of your office. The cabinet door is locked, so you fumble in your pocket for a key. Nearby, a stand holds sheet music and a guitar. On your coffee table is an odd assortment of magazines, from *Wired* to *Sushi News, Chess Life,* and *Scientific American.*

Around the table are three antique Chippendale chairs, covered in tapestry. Sally takes a seat and stares at the paintings on the wall—various French Postimpressionist works with lush colors, especially those by Paul Gauguin. Your favorite is *Spirit of the Dead Watching* depicting a woman lying on a bed with an owl flying above her and a man, dressed in black, sitting nearby. The painting is an excellent reproduction—the original is worth millions of dollars, hardly affordable on the meager salary the FBI pays you.

You unlock the cabinet and withdraw a large bone. "Sally, take this bone."

"Where did you get this thing?"

"It's the longest human bone on record—a 29.9-inch thigh bone of the German giant Constantine, who died in Mons, Belgium, in 1902 at age thirty."

Sally eyes you suspiciously.

"Suppose I put a pin through one end of this bone and rotated the bone on a table. What shape would its free end trace out?"

"A circle."

"Now start swinging the bone all around, but keep your hand in one point as much as possible."

Sally starts swinging the bone in all directions. "What are you getting at? My arm is getting tired."

"Now the free end is tracing out a sphere. Assume now that space has a fourth coordinate at right angles to the other three and that you could swing the bone in four-space. The free end would generate a hypersphere.

"The surface of an *n*-sphere has a dimensionality of $n - 1$. For example, a circle's 'surface' is a line of one dimension. A sphere's surface is two-dimensional. A hypersphere's surface is three-dimensional. Many physicists in the late 1800s thought forces like gravity and electromagnetism could be transmitted by vibrations of a hypersphere."

Sally tosses the bone to you. "You already told me Einstein proposed the surface of a hypersphere as a model of our universe. It would be finite but boundaryless, like the surface of a ball. It's an interesting theory."

You catch the bone with a deft flick of your wrist and toss it onto a chair.

You reach into a cabinet and remove two tennis balls.

Sally stares at your balls. "Why is it so hard to imagine our space as hyperspherical?"

You throw one of the balls to Sally. "The curvature of our 3-D universe would be in the direction of the fourth dimension. Our 'straight lines' would actually be curved, but in a direction unknown to us. This would be similar to a creature living on the two-space surface of a sphere. Lines that appeared straight to him would actually be curved. Parallel lines could actually intersect, just as longitude lines (which seem parallel at the equator) intersect at the poles. This curvature could be hard to detect if his, or our, universe were large compared to the local curvature. In other words, only if the radius of the hypersphere (whose hypersurface forms our 3-D space) were very small, could we notice it."

Sally plays with the ball, studying its smooth surface. "What would happen if we lived in a hypersurface of a hypersphere whose radius was the size of a football stadium?"

"In such a small universe, if you run in a straight line, you'd return to your starting point very quickly. In any direction you looked, you'd see yourself" (Fig. 4.2a). You pause dramatically before launching into a more intriguing line of thought. "The idea that our 3-D space is the surface of the hypersphere is seriously considered by many responsible scientists. This idea suggests another, even wilder possibility."

"Yes?"

"Sally, consider Flatland existing as a surface of a sphere. Pretend the surface of the tennis ball in your hand is Flatland. Three dimensions permit the possibility of many separate, spherical Flatlands floating in 3-D space. Think of many floating bubbles in which each bubble's surface is an entire universe for Flatlanders. Similarly, there could be many hyperspherical universes floating in 4-D space" (Fig. 4.2b).

Sally nods. "If there are many hyperspherical universes, why can't we escape from our hypersphere and explore these other universes?"

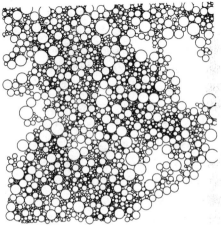

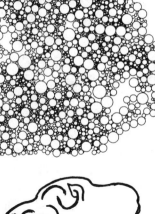

Figure 4.2 Strange universes. (a) Tapping yourself on the back in a small, closed universe. (Drawing by Clay Fried.) (b) Like bubbles floating in the air, many separate n-dimensional worlds could exist if the universe were $n+1$-dimensional. Could our universe be one of many separated in 4-D space? If these spaces were to touch at some point, would we be able to communicate with an adjacent world? (c) Your "inside-out" friend in a small, hyperspherical universe. (Drawing by Clay Fried.)

A

B

C

"That's a good question.[1] I have no answer, but now it seems that the Omegamorphs are changing all that."

Sally tosses the tennis ball to you. "If we lived on a hyperspherical universe and you traveled to a point in the universe that was furthest from me, I would still be able to see you with a telescope in no matter what direction I looked."

"That's true if the sphere were not too large."

"If I could see you no matter where I looked, would you seem infinitely big to me?"

You place the tennis balls back in the cabinet. "Yes. Even stranger, I would be 'inside out.' In other words, instead of my skin and hair forming a surface around my guts, it would seem that my guts were on the outside, and my hair and skin would be surrounding you! Of course, I wouldn't notice anything strange. I wouldn't be about to die. To me, *you* would look infinitely large and inside out" (Fig. 4.2c).

You toss a basketball to Sally. "Next lesson. If you cut this sphere with a plane, you'd produce a circle. If you cut a hypersphere with a 3-D hyperplane, the cross section is a sphere."

Sally looks at the basketball. "What happens if you try to cut a hypersphere with an ordinary 2-D plane?"

"Sally, you can't slice a hypersphere into two pieces with a 2-D plane. A hyperbasketball, sliced down the middle by a plane, remains in one piece, just like a sphere pierced with a line does not fall apart into two separate pieces. This means that a guillotine for a hyperbeing would be a 3-D object like a cube, not a plane."

You take the basketball from Sally's hands and deflate it with a pin. "If I asked you to turn this basketball inside out, could you?"

Sally studies it for a moment. "I don't think so, not without cutting it."

You nod. "Correct. However, a flexible sphere of any dimension can be turned inside out through the next-highest dimension. For example, we 3-D beings can turn a rubber ring inside out so that its outer surface becomes the inner, and the inner becomes the outer. Try it with a rubber band. Similarly, a hyperbeing could grab this basketball and turn it inside out through his space."

"Does this mean that a hyperbeing could turn a human inside out?"

"From a practical standpoint, we're not quite as flexible as a rubber ball. We're also not spheres. We're more like a sphere with a digestive tube running down the middle. But you're right, topologically speaking, a hyperbeing could do weird things to us."

Outside your window, you see the man in the Santa Claus outfit. "Who is that?"

Sally looks out. "No one special, I'm sure."

You try desperately to glimpse his face, but there is not enough light. All you can see is a figure dressed in a red suit. Even though you can't observe the man's face, you recognize something familiar. On his left hand is a tattoo in the shape of a tesseract projected into two dimensions. Could he be an agent of the Omegamorphs? Worse, you sense that the man is looking for you.

You are paralyzed; you are certain the Santa Claus man knows you are there. Your body tenses, waiting—but for what? Outside on the street, the sounds blend in a cacophonous hiss. You hear voices, but can never identify sentences. There is some laughter.

You blink and the man is gone. Just as many people are walking by on the sidewalk, but the sounds are softer, less tense.

Sally taps you on your back. "It was no one. It's that time of year."

You nod and withdraw a wooden cube from your cabinet. "Follow me to the blackboard. I want to talk about tesseracts, the 4-D analogs of a cube. You can get an idea about what they're like by starting in lower dimensions. For example, if you move a point from left to right you trace out a 1-D line segment." You place the tip of your chalk on the blackboard and move the tip to the right so that it produces a line. "If you take this line segment and move it up (perpendicularly) along the blackboard, you produce a 2-D square. If you move the square out of the blackboard, you produces a 3-D cube" (Fig. 4.3).

Sally comes closer. "How can we move the square out of the black board?"

"We can't do that, but we can graphically represent the perpendicular motion by moving the square—on the blackboard—in a direction *diagonal* to the first two motions. In fact, if we use the *other* diagonal direction to represent the fourth dimension, we can move the image of the cube in this fourth dimension to draw a picture of a 4-D hypercube, also known as a tesseract. Or we can rotate the cube and move it straight up in the drawing" (Fig. 4.4).

"Beautiful. The tesseract is produced by the trail of a cube moving into the fourth dimension."

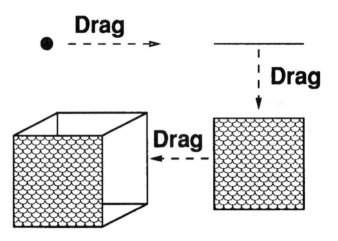

Figure 4.3 Lower-dimensional figures trace out higher-dimensional figures when the lower-dimensional figures are moved.

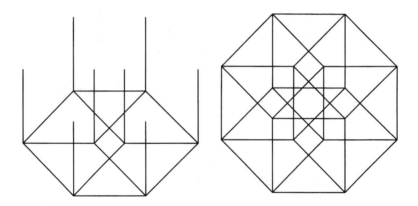

Figure 4.4 A hypercube (right) produced by moving (left) a cube along the fourth dimension.

"That's right, Sally. Our visual powers have a hard time moving the image of a cube, but we can assume that the cube is shifted a distance in a direction perpendicular to all three of its axes. We can even write down the number of corners, edges, faces, and solids for higher-dimensional objects." You write on the blackboard:

	Corners	Edges	Faces	Solids	Hyper-volumes
Point	1	0	0	0	0
Line segment	2	1	0	0	0
Square	4	4	1	0	0
Cube	8	12	6	1	0
Hypercube	16	32	24	8	1
Hyperhypercube	32	80	80	40	10

You stare into Sally's eyes with pupils slightly dilated in the dim room light. "Take a look at the hypercube drawing. Can you see the sixteen corners? The number of corners (or vertices) doubles each time we increase the dimension of the object. The hypercube has thirty-two edges. To get the volumes of each object, all you have to do is multiply the length of the sides. For example, the volume of a cube is l^3 where l is the length of a side. The hypervolume of a hypercube is l^4. The hyper-hypervolume of a 5-D cube is l^5, and so on.

"How can we understand that a hypercube has thirty-two edges?"

"The hypercube can be created by displacing a cube in the upsilon or delta direction and seeing the trail it leaves. Let's sum the edges. The initially placed cube and the finally placed cube each have twelve edges. The cube's eight corners each trace out an edge during the motion. This gives a total of thirty-two edges. The drawing is a nonperspective drawing, because the various faces don't get smaller the 'further' they are from your eye" (Fig. 4.4).

You hand Sally a cube of sugar and a pin. "Can you touch any point inside any of the square faces without the pins going through any other point on the face?"

"Of course."

"Sally, let's think what that would mean for a hyperman touching the cubical 'faces' of a tesseract. For one thing, a hyperman can touch any point inside any cubical face without the pin's passing through any point in the cube. Points are 'inside' a cube only to you and me. To a hyperman, every point in each cubical face of a tesseract is directly exposed to his vision as he turns the tesseract in his hyperhands."

You go to the blackboard and begin to sketch. "There's another way to draw a hypercube. Notice that if you look at a wire-frame model of a cube with its face directly in front of you, you will see a square within a

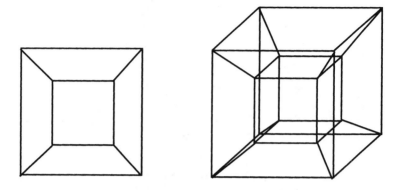

Figure 4.5 A wire-frame model of a cube viewed head-on and a tesseract.

square. The smaller square is further away from your eye and is drawn smaller because the drawing is a perspective drawing. If you looked at a hypercube in the same manner, you would see a cube within a cube. The closest part of the hypercube appears as a large cube, and the part farthest away appears as a smaller cube inside the larger one. This is called a 'central projection' of the hypercube. More accurately, it is a plane projection of a 3-D model that is in turn a projection of a hypercube (Fig. 4.5). This is a shadow you might see if a hypercube is illuminated from a point upsilon above ordinary space in the fourth dimension."

Sally studies the figure. "A cube is bounded by square faces and a hypercube by cubical faces?"

"Right."

"A hypercube contains eight cubes on its hypersurface?"

"Correct."

"But I don't see them in your central-projection drawing" (Fig. 4.5).

"Sally, six of the eight cubes are distorted by projection just as four of the cube's square faces are distorted when drawn on a plane. For a tesseract, the eight cubes are: the large cube, the small interior cube, and six hexahedrons (distorted cubes) surrounding the small interior cube."

You smile as you look into Sally's upturned face and study the paleness of her skin, the smoothness of her lips, the soft wetness of her eyes. She has the eyes of a doe. For an instant you imagine yourself and Sally in front of a roaring fireplace with slender glasses of champagne. Ah, but such a fantasy is ridiculous. You are professionals.

"Sally, ever wonder what a hypersphere would look like projected into our universe?"

She smiles. "Sure, every day, every waking hour."

You bring out a globe made of glass with all the continents marked. You shine a light on it and look at the projection on the wall. "First, let's consider the projection of an ordinary sphere onto a plane." You point at a spot projected on the wall. "Notice that the two hemispheres will overlap on one another, and that the distance between our FBI headquarters and China seems very short. Of course, that's only because we're looking at a projection. In fact, every point on the projection represents two opposite points on the original globe. China and America don't actually overlap because they are on opposite sides of the globe."

Sally studies the projection on the wall. "What we're seeing looks like two flat discs put together and joined along their outer circumferences."

"Right. Watch as I rotate the globe. The projection makes it appear that the Earth is rotating both right and left simultaneously. Would a 2-D being go insane trying to picture an object rotating in three dimensions?"

"Imagine our difficulty in visualizing a 4-D rotating planet!"

You nod. "You can imagine a space-projection of a hypersphere into our world as two spherical bodies put through each other and joined along their outer surfaces. It would be like two apples grown together in the same regions of space and joining at their skins."

You place the globe on an old Oriental carpet covering the hardwood floor of your office. "Let's return our attention to hypercubes. Another way to represent a hypercube is to show what it might look like if it was unfolded." You bring out a paper cube that has been taped together and remove some pieces of the tape. "By analogy, you can unfold the faces of a paper cube and make it flat" (Fig. 4.6).

You then bring out a paper model of an unfolded hypercube. "Sally, we can cut a hypercube and 'flatten' it to the third dimension in the same way we flattened a cube by unfolding it into the second dimension. In the case of the hypercube, the 'faces' are really cubes" (Fig. 4.7).

You point at a poster on the wall. "The hypercube has often been used in art. My favorite is the unfolded hypercube from Salvador Dali's 1954 painting *Corpus Hypercubus* (Fig. 4.8). By making the cross an unfolded tesseract, Dali represents the orthodox Christian belief that Christ's death was a metahistorical event, taking place in a region outside of our space

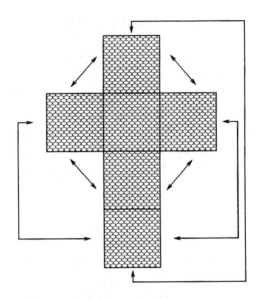

Figure 4.6 One way to unfold a cube. The arrows show a way of folding the faces to reform the cube—for example, the bottom face connects to the top face.

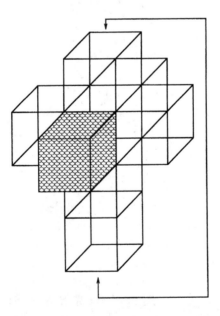

Figure 4.7 One way to unfold a hypercube. Just as with the cube in Figure 4.6, the bottom cubical "face" must join with the top "face" when folding the cubes to reform the hypercube. This folding must be done in the fourth dimension. (The forwardmost cubical face is shaded to help clarify the drawing.)

Figure 4.8 *The Crucifixion* (*Corpus Hypercubus*), by Salvador Dali (1954). Oil on canvas, Metropolitan Museum of Art, Chester Dale Collection. © 1999 Artists Rights Society, New York.

and time. We ordinary humans could only see the event with limited vision on an unfolded tesseract."

Sally holds the unfolded hypercube in her slender hands. "We've talked about how a cubical prison couldn't trap a 4-D being. But could a tesseract prison hold such a creature? Perhaps we could trap the creature that took the president in a hypercube."

You nod. "That's correct."

"What would a hypercubical prison look like in our world?"

"Sally, again let's reason by analogy and consider what a cubical prison would look like to a Flatlander. Let's pretend that the prison is a hollow cube made of steel. As the prison was pushed down through Flatland, the Flatlander would first see a solid square face. This is the floor of the prison. Next, the walls would be pushed down, forming a hollow square on Flatland. Finally, the Flatlander would see a solid square face corresponding to the jail's ceiling. If the president were in this cubical jail while it was pushed down into Flatland, we'd first see cross sections of his feet, then body, then head, until he disappeared."

"If we pushed down the cubical prison at odd angles, we might see other intersections with Flatland."

You motion Sally over to the water tub in your office—the one you use for understanding intersections of 3-D objects with Flatland. "Sally, you're right as usual. In my example, I pushed the prison at right angles so it had a square cross section. But if we tip the prison so that one of its corners faces down, we'd first see a single point, then a triangle, then a six-sided figure (a hexagon), then a triangle, and finally a point." You slowly push a glass cube into the water, corner first, to show Sally the various cross sections (Fig. 4.9).

"Now let's consider a *hypercube* prison containing the president and also the Omegamorph that abducted him. If the hypercube were pushed delta into our space, we might first see its 'cubical floor.' This floor would be a solid cube of steel corresponding to the steel face of the 3-D prison. Next we'd encounter hollow steel cubes, and finally the solid steel cube 'ceiling.' If the cube were made of glass so we could see inside, the president might materialize all at once in the same way that a Flatlander aligned parallel to Flatland might materialize all at once as he intersected Flatland."

"The 4-D Omegamorph would look like hair or skin blobs as the 4-D prison was lowered into our world, and the tesseract intersecting our world could look like an ordinary hollow cube."

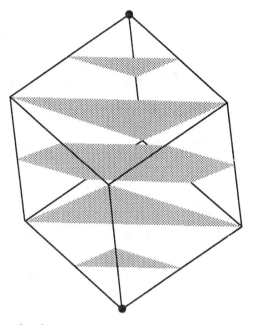

Figure 4.9 As the cubical prison moves down, corner-first, through Flatland, the Flatlanders first see a point, then a series of growing triangles, then hexagons, then triangles, then a point. (After Banchoff.)

You nod. "Sally, if Satan were a 4-D being, it might be possible to confine him in a tesseract prison. If it were an ordinary cube, Satan could flee into the fourth dimension. However, if the cube in our world was really part of a hypercube, he would be trapped. As he leaped upsilon into the fourth dimension, he'd just hit his head on a cubical ceiling. Similarly, a Flatlander having the ability to leap into the third dimension would bang himself on a cubical prison that spanned his world."

You and Sally are quiet as you gaze outside at the street lights. They cast shadows that spring up about you as if they are living creatures.

Suddenly most of the lights go out. Others flicker like fireflies.

You turn to Sally. "Must be an electrical problem."

She comes closer to you. "Spooky."

I want to fly, you think as you stare into the dim room. Fly into the fourth dimension, like a bird. To fly, to feel the surge of your body as it lifts into the higher universe. Breathless.

You imagine a force that moves you gently but purposefully upward. Voices float around you like the breaking of ocean waves on rocks.

You look out the window, occasionally seeing fantastically bright birds alighting on the hoods of cars. Could they be birds from the fourth dimension? Perhaps "bird" is not the right word. Their gossamer wings quiver on bodies resembling balls of twine. Some of the animals have more slender bodies. Speckles on their skin glow like neon lights. Then the speckles fade in the evening drizzle.

The only illumination comes from the green and red lights emitted by the bioluminescent bacteria coating the trees. It reminds you of Christmas.

The Science Behind the Science Fiction

The rift between belief and natural science can today be bridged only if it is possible to transpose the concept of space, which has acquired a position of primary significance in modern physics, in a higher connotation to the world-picture of belief.

— Karl Heim, *Christian Faith and Natural Science*

Theoretical physics seems to becoming more and more like science fiction.

— Steven Weinberg, *The First Three Minutes*

Hinton Cubes

One of the greatest challenges in understanding the fourth dimension is visualizing 4-D objects. Nineteenth-century German physicist Hermann von Helmholtz believed that the human brain could visualize the fourth dimension if it had the correct input data. Helmholtz's detailed investigation of vision led him to refute Immanuel Kant's theory that space is a fixed absolute thing with a reality of its own, independent of material objects (see Kant's 1768 paper "On the First Ground of the Distinction of Regions in Space"). Helmholtz tried to show exactly how the sense of vision created the idea of space. In other words, Helmholtz believed that space was a learned, not an inherent, concept. Moreover, Helmholtz also attacked Kant's insistence that space had to be three-dimensional because that was how the mind had to conceive it. For example, Helmholtz used his considerable mathematical talents to investigate the properties of non-Euclidean space and showed that it could be conceived and worked with almost as easily as the geometry of three dimensions.

Kant's first published paper, "Thoughts on the True Estimation of Living Forces" (1747), suggests that he was curious about the fourth dimension. In his paper, Kant asks, "Why is our space three-dimensional?" He uses physics to remind us that forces like gravity seem to move through space like expanding spheres; that is, their strength various inversely with the square of the distance. Kant reasoned that if God chose to make a world where forces varied inversely with the *cube* of the distance, God would have required a space of four dimensions.

In the late 1800s, English mathematician Charles Howard Hinton spent years creating new methods by which ordinary people could "see" 4-D objects.[2] Eventually, he invented special cubes that were said to help visualize hypercubes. These models would come to be known as Hinton cubes and were advertised in magazines and even used in seances. By meditating on Hinton cubes, it was rumored that people could not only catch glimpses of the fourth dimension but also ghosts of dead family members.

Charles Hinton studied mathematics at Oxford, married Mary Boole (one of the daughters of famous logician Geoge Boole), and then moved to the United States after being convicted of bigamy. He taught mathematics at Princeton University and the University of Minnesota. In 1907, Hinton published *An Episode of Flatland* (a work more scientific than Abbott's *Flatland*) in which 2-D creatures resided on the surface of a circular world called Astria. Gravity behaves as it does in our world, except that on the plane its force varies inversely with distance instead of with the square of distance.

In Hinton's book, Astrians have only one eye, just as Abbott's Flatland creatures. (In principle, both authors could have given their creatures two eyes, each with 1-D retinas, to provide binocular vision.) To pass each other as they travel on the surface of Astria, the inhabitants must go under or over each other, like acrobats. All female Astrians are born facing west; all males are born facing east. Astrians keep their orientation until they die because there is no way to "flip over" without being rotated in the third dimension. To kiss his son, an Astrian dad must hold the boy upside down! (It's too bad that Astrians didn't have long necks that would permit them to tilt their heads backwards and upside down to see behind them.)

What would it be like to live on Astria, a fully developed 2-D world with gravity and all the laws of physics? For one thing, it would be difficult to build houses that have several windows open at the same time. For example, when the front window is open, the window in the back must be kept closed to keep the house from collapsing. Perfectly hollow tubes and pipes would be difficult to construct. How would you keep both sides of the pipes together without sealing the tube? It might be possible to have tubes with a series of interlocking valves, like the self-gripping gut discussed in Chapter 2. You could make a tunnel with a series of doors that closed behind you as you walked. But you could never have all doors open at once or the tunnel could collapse. Ropes could not be knotted since line segments don't knot in two-space. Hooks and levers would work just fine. Birds could still fly by flapping their wings.

In Hinton's book, one of the Astrians comes to realize that there is a third dimension and that all Astrian objects have a slight 3-D thickness. He believes that the Astrians slide about over the smooth surface of what he calls an "alongside being." In a moving speech to his fellow Astrians, he proclaims:

> Existence itself stretches illimitable, profound, on both sides of that alongside being. . . . Realize this . . . and never again will you gaze into the blue arch of the sky without added sense of mystery. However far in those never-ending depths you cast your vision, it does but glide alongside an existence stretching profound in a direction you know not of.
>
> And knowing this, something of the old sense of the wonder of the heavens comes to us, for no longer do constellations fill all space with an endless repetition of sameness, but here is the possibility of a sudden and wonderful apprehension of beings, such as those of old time dreamed of, could we but . . . know that which lies each side of all the visible.

Next time there is a bright blue arch of sky above, gaze at it and recall the words of Charles Hinton.

Let us return for a moment to Hinton's cubes. Hinton's methods of visualizing four-space structures in three-space cross sections required hundreds of small cubes, colored and labeled. Hinton said that he was able to think in four dimensions as a result of studying his cubes for years. He also noted he taught the method to his sister-in-law when she was eighteen. Although the girl had no formal training in mathematics, she soon developed a remarkable grasp of the 4-D geometry and later made significant discoveries in the field.

Hinton's disciples spent days mediating on the cubes until some thought they could mentally reassemble these cubes in the fourth dimension—thus achieving nirvana. Figure 4.7 shows an unraveled hypercube. Although the cubes of this tesseract seem static, a 4-D person can fold the cubes into a hypercube by lifting each individual cube off our universe into the fourth dimension. Note that Hinton used the words "ana" and "kata" in the same way I use the terms "upsilon" and "delta" to describe motions in the 4-D world as counterparts for terms like "up" and "down." (I find that upsilon and delta are easier to remember than ana and kata because of the "up" in upsilon and "d" in delta.)

Unraveling

Take a deep breath, and let your imagination soar. Watch now as a 4-D person folds a tesseract into a hypercube. What do you see? Not much! All you observe are the various cubes in Figure 4.7 disappearing, leaving only the center cube in our universe. The folded hypercube looks just like an ordinary cube in the same way a cube can appear like an ordinary square to a Flatlander.

What would it be like to be visited by a hypercube? If it came into our universe "cube-first" (like a cube coming into a planar universe "face first"), we would just see a cube that disappeared as it finally went through our 3-D world. Even though you and I are not likely to be able to "see" a hypercube all at once in the same way that we can see a cube, we can be sure that such an object would have sixteen vertices. It might even look like just a square when it just touched our world. However, if the object rotated, the ordinary-looking square could reveal a starburst of lines (as in Fig. 4.4) corresponding to an object that really has twenty-four square faces, thirty-two edges, and sixteen vertices. If a 5-D cube passed through a 4-D universe, it would appear for a while as a hypercube with thirty-two vertices before it disappeared entirely from the world.[3]

In this chapter we've discussed how a cube would appear to Flatlanders as it penetrates their world. As a cube moves corner-first through Flatland, the Flatlanders first see a point, then a series of growing triangles, then hexagons, then triangles, and finally a point (Fig. 4.9). Similarly, a variety of shapes would be produced as a hypercube penetrated our world (Fig. 4.10). In four dimensions, slicing with a 3-D knife produces an array of strange shapes ranging from distorted cubes, various prisms, and polyhedra often in unexpected arrangements.

What would the hypercube look like as we rotated it? One way to visualize this is to begin with the "central projection" in Figure 4.5 and watch the wireframe change as the hypercube rotates (Fig. 4.11). Starting from the left and going clockwise, the large exterior cube opens toward the top, flattens out, and opens inward to form an incomplete pyramid. Meanwhile, the smaller interior shaded cube opens toward the bottom to form another incomplete pyramid. If we continue to rotate the hypercube, the unshaded exterior cube will become the small cube, and the small cube will flatten out and come back to become the large cube. As Thomas Banchoff notes, each of the eight cubical faces takes its turn holding all the various positions in this projection. As each of the cubes flattens out and opens up again during the rotation, it changes orientations. If a cube contained a right-handed object before the flattening, the object would become left-handed afterwards.

On the Trail of the Tesseract

In this chapter, we've also used lower-dimensional analogies to help contemplate the mathematics of higher spaces. Throughout history, mathematicians have used interdimensional analogies. For example, if mathematicians understood a theorem in plane geometry, they were often able to find analogous theorems in solid geometry. (Theorems about circles provide insight into theorems about spheres and cylinders.) Similarly, solid geometry theorems have suggested new relationships among plane figures. If history shows that knowledge can be gained by going to higher dimensions, imagine what we might learn by contemplating 4-D geometries.

The most often-used analogy for contemplating shapes in higher dimensions involves moving objects perpendicular to themselves. If we move a 0-D point with no degrees of freedom, we generate a line, a 1-D object with two end points (Fig. 4.3). A line moved perpendicular to itself along a plane generates a square with four corners. A square moved perpendicular to itself forms a cube

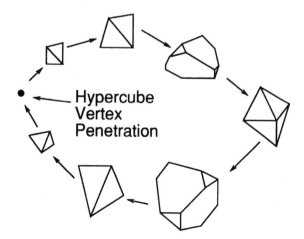

Figure 4.10 Slices of a hypercube as it moves corner-first through our world.

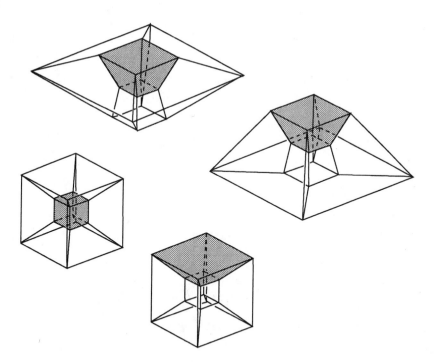

Figure 4.11 Central projections of a rotating hypercube in four-space. Starting from the left and going clockwise, the shade cube opens toward the bottom; if the rotation is continued, the small cube will flatten out and come back to become the large exterior cube. (After Banchoff.)

with eight corners. Even though we cannot easily visualize the next step in the process, we can predict that if we were able to move a cube perpendicular to all its edges, we would generate a 4-D object: a hypercube. It would have sixteen corners. The trend in the number of corners is a geometric progression (2, 4, 8, 16 . . .), and we can therefore calculate the number of corners in any dimension by using the formula 2^n where n is the number of dimensions.

We can also consider the number of boundaries for objects in different dimensions. A line segment has two boundary points. A square is bounded by four line segments. A cube is bounded by six squares. Following this trend, we would expect a hypercube to be bounded by eight cubes. This sequence follows an arithmetic progression (2, 4, 6, 8 . . .).

The area of a square of edge length a is a^2. The volume of a cube of edge length a is a^3. The hypervolume of an n-cube is a^n.

Past books typically provide wire-frame diagrams for tesseracts produced by the "trail" of a cube as it moves in a perpendicular direction, similar to the one in Figure 4.4. Of course, we can't really move in a perpendicular direction, but we can move the cube diagonally, in the same way a square is moved diagonally to represent a cube. Now prepare yourself for some wild trails of higher-dimensional objects rarely, if ever, seen in popular books. To give you an idea of the beauty and complexity of higher-dimensional objects, I produced Figures 4.12 to 4.17 using a computer program. Modern graphics computers are ideal tools for visualizing structures in higher dimensions.

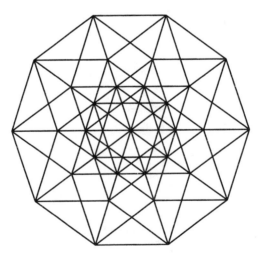

Figure 4.12 A 5-D cube produced by moving a hypercube along the fifth dimension.

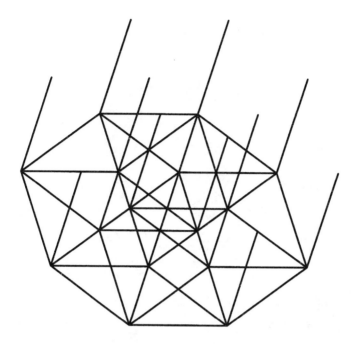

Figure 4.13 Embryonic 5-D cube in Figure 4.12 prior to the dragging of the 4-D cube.

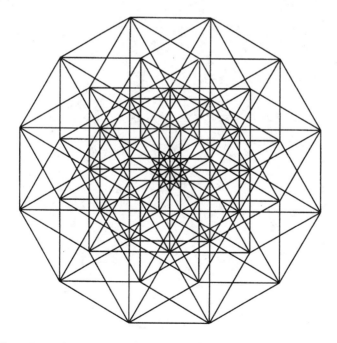

Figure 4.14 A 6-D cube produced by moving a 5-D cube along the sixth dimension.

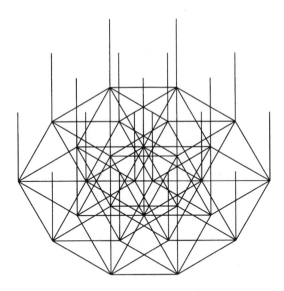

Figure 4.15   Embryonic 6-D cube in Figure 4.14.

Figure 4.16   A 7-D cube produced by moving a 6-D along the sixth dimension.

Figure 4.17   An 8-D cube.

We can think of these representations as "shadows" of hypercubes on 2-D pieces of paper. Luckily, we don't have to build the object to compute what its shadow would look like. (The computer code I used to create these forms is listed in Appendix I.) Projections of higher-dimensional worlds have stimulated many traditional artists to produce geometrical representations with startling symmetries and complexities (Figs. 4.18 to 4.20).

Although computer graphic devices produce projections of higher dimensions on mere 2-D screens, the computer can store the location of points in higher dimensions for manipulations such as rotation and magnification. The computer can then display projections of these higher-dimensional forms from various viewpoints. In fact, the computer is frequently used to represent higher dimen-

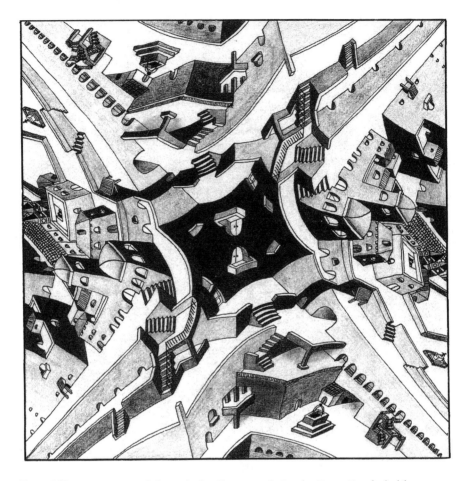

Figure 4.18 Symmetrical, hyperbolically warped city, by Peter Raedschelders.

sions in all kinds of practical scientific problems where temperature or electric charge are the additional "dimensions" represented as colors on 3-D objects.

We can prepare ourselves for any invasions of 4-D creatures entering our world. Although it may be difficult for us to fully *see* higher dimensions, we can use computers to develop ways of responding to the intersections of higher-dimensional phenomena in our world. The computer also makes us gaze in awe at the beauty and complexity of higher dimensions. On this theme, Professor Thomas Banchoff of Brown University writes

> The challenge of modern computer graphics fits right in with one of
> the chief aims of Edwin Abbott Abbott in the introduction of his time-

Figure 4.19 Fish in higher dimensions, by Peter Raedschelders.

less book, namely to encourage in the races of solid humanity that estimable and rare virtue of humility. We will continue to appreciate *Flatland* more and more in the years to come.

Distance

Many readers will be familiar with how to compute the distance *d* between two points (x_1, y_1) and (x_2, y_2) on a plane:

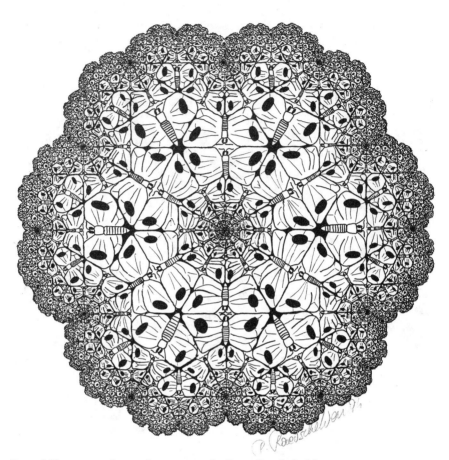

Figure 4.20 Butterflies in hyperspace, by Peter Raedschelders.

$$d = \sqrt{(x_1 - x_2)^2 + (y_1 - y_2)^2}$$

(You can derive this equation by drawing diagrams and using the Pythagorean theorem, which states that the length of the hypotenuse of a right triangle equals the square root of the sum of the squares of the other two sides.) Simply by adding another term, this formula can be extended so that we can compute the distance between two points in three dimensions:

$$d = \sqrt{(x_1 - x_2)^2 + (y_1 - y_2)^2 + (z_1 - z_2)^2}$$

Similarly, we may extend this formula to 4, 5, 6, . . . or k dimensions! Various scholars have debated whether humans can truly grasp the meaning of a

4-D line or 4-D distance. Edward Kasner and James Neuman remarked in 1940 (the same year that Heinlein published his science-fiction tale about the 4-D house):

> Distance in four dimensions means nothing to the layman. Even four-dimensional space is wholly beyond ordinary imagination. But the mathematician is not called upon to struggle with the bounds of imagination, but only with the limitations of his logical faculties.

Hyperspheres

> We sail within a vast sphere, ever drifting in uncertainty, driven from end to end.
>
> —Blaise Pascal, *Pensées*

I would like to delve further into the fourth dimension by discussing hyperspheres in greater detail. Let's start by considering some exciting experiments you can conduct using a pencil and paper or calculator. My favorite 4-D object is not the hypercube but rather its close cousin, the *hypersphere*. Just as a circle of radius r can be define by the equation $x^2 + y^2 = r^2$, and a sphere can be defined by $x^2 + y^2 + z^2 = r^2$, a hypersphere in four dimensions can be defined simply by adding a fourth term: $x^2 + y^2 + z^2 + w^2 = r^2$, where w is the fourth dimension. I want to make it easy for you to experiment with the exotic properties of hyperspheres by giving you the equation for their volume. (Derivations for the following formulas are in the Apostol reference in Further Readings.) The formulas permit you to compute the volume of a sphere of *any* dimension, and you'll find that it's relatively easy to implement these formulas using a computer or hand calculator. The volume of a k-dimensional sphere is

$$V = \frac{\pi^{k/2} r^k}{(k/2)!}$$

for *even* dimensions k.

The exclamation point is the mathematical symbol for factorial. (Factorial is the product of all the positive integers from one to a given number. For example, $5! = 1 \times 2 \times 3 \times 4 \times 5 = 120$.) The volume of a 6-D sphere of radius 1 is

$\pi^3/3!$, which is roughly equal to 5.1. For odd dimensions, the formula is just a bit more intricate:

$$V = \frac{\pi^{(k-1)/2} m! 2^{k+1} r^k}{(k+1)!}$$

where $m = (k+1)/2$.

The formulas are really not too difficult to use. In fact, with these handy formulas, you can compute the volume for a 6-D sphere just as easily as for a 4-D one. "Code 2" in Appendix F lists some of the computer program steps used to evaluate this formula.

Figure 4.21 plots the volume of a sphere of radius 2 as a function of dimension. For radius 2 and dimension 2, the previous equations yield the value 12.56, which is the area of a circle. A sphere of radius 2 has a volume of 33.51. A 4-D hypersphere of radius 2 has a volume of 78.95. Intuitively, one might think that the volume should continue to rise as the number of dimensions increase. The volume—perhaps we should use the term "hypervolume"—does grow larger and larger until it reaches a maximum—at which point the radius 2 sphere is in the twenty-fourth dimension. At dimensions higher than 24, the volume of this sphere begins to decrease gradually to zero as the value for dimension increases. An 80-D sphere has a volume of only 0.0001. This apparent turnaround point occurs at different dimensions depending on the sphere's radius, r.

Figure 4.22 illustrates this complicated feature by showing volume plots of a k-dimensional sphere for radius 1, 1.1, 1.2, 1.3, 1.4, 1.5, and 1.6 as a function of dimension. For all the sphere radii tested, the sphere initially grows in volume and then begins to decline. (Is this true for all radii?) For example, for $r = 1$, the maximum hypervolume occurs in the fifth dimension. For $r = 1.1$, the peak hypervolume occurs in the seventh dimension. For $r = 1.2$, it occurs in the eighth dimension. (Incidentally, the hypersurface of a unit hypersphere reaches a maximum in the seventh dimension, and then decreases toward zero as the dimension increases.) Here is a great example of how simple graphics, like the illustration in Figure 4.22, help us grasp the very nonintuitive results of a hypergeometrical problem! If we examine the equations for volume more closely, we notice that this funny behavior shouldn't surprise us too much. The denominator contains a factorial term that grows much more quickly than any power, so we get the curious result that an infinite dimensional sphere has no volume.

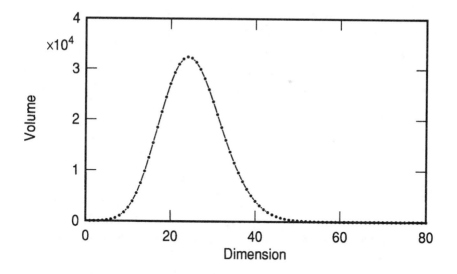

Figure 4.21   Volume of a radius 2 sphere as a function of dimension.

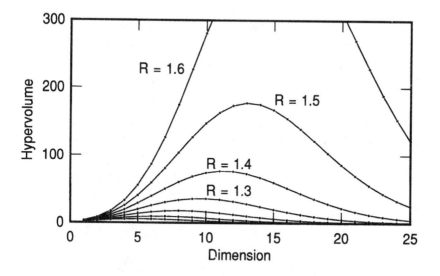

Figure 4.22   Volume of K-dimensional spheres for radii of 1, 1.1, 1.2, 1.3, 1.4, 1.5, and 1.6.

Using the equations for volume given here, you'll find that an 11-D sphere of radius 2 feet is 333,763 feet.[11] Considering that the volume of a brontosaurus (nowadays called an apatosaurus) is about 100,000 cubic feet, does this mean that the brontosaurus could be crammed into this small 11-D sphere? This amusing thought is just a prelude to the questions that follow.

Some Hypersphere Questions

> She reappeared, looking back at him from her fat flat suspicious face, and Kevin understood the reason why she had disappeared for a moment. It was because the concept of "a side view" didn't exist in a world where everything was perfectly flat. This is Polaroidsville, he thought with a relief which was strangely mingled with horror.
> —Stephen King, *Four Past Midnight*

Now the time has come for considering some really tough questions. If you are a teacher, why not give these to your students to answer.

1. Examine the graph in Figure 4.21. Could a 24-D sphere of radius 2 inches contain the volume of a blue whale (Fig. 4.23)?
2. Could a 1000-D sphere of radius 2 inches contain the volume of a whale, considering that the sphere's hypervolume is very, very close to 0 (Fig. 4.21)?
3. Could a Sea World animal trainer fit a whale into an 8-D sphere (of radius 1 inch) as its aperture intersected with our 3-D world?
4. The number of atoms in a human's breath is about 10^{21}. If each atom in the breath were enlarged to the size of a marble, what percentage of a human's breath could fit into a 16-D hypersphere of radius 1.1 inches?
5. Estimate the value of the 24-D hypervolume of a whale. To compute this, assume that the length of a blue whale is about 100 feet.
6. What is the one-million-dimensional hypervolume of the earth? Assume the earth to have a diameter of 4.18×10^7 feet. Also, can you approximate the 4-D hypervolume of Albert Einstein's brain? (The brain of an average adult male weighs 3 lbs. 2.2 oz., decreasing gradually to 3 lbs. 1.1 oz with advancing age.) Another problem: In 1853, the 350-foot-tall Latting Observatory in New York City was the highest manmade structure in North America (Fig. 4.24). Estimate its 4-D hypervolume.

How to Stuff a Whale into a Five-Dimensional Sphere

The answers to the previous six questions are: yes, yes, no, 100 percent, zero, and zero (for all parts of question 6). To help understand these answers, consider the act of stuffing rigid circular regions of a plane into a sphere. If the circular discs are really two-dimensional, they have no thickness or volume. Therefore, in theory, you could fit an infinite number of these circles into a sphere—provided that the sphere's radius is slightly bigger than the circle's radius. If the sphere's radius were smaller, even one circle could not fit within the enclosed volume since it would poke out of the volume. Therefore, in answer to question 1, the volume of a whale *could* reside comfortably in a 24-D sphere with a radius of 2 inches. In fact, an infinite number of whale volumes could fit in a 24-D sphere. Likewise, in answer to question 2, a 1000-D sphere with a radius of 2 inches could contain a volume equivalent to that of a whale. However, you could not physically stuff a whale into either of these spheres because the whale has a minimum length that will not permit it to fit. (Con-

Figure 4.23   A whale waiting to be stuffed into a 24-D sphere.

Figure 4.24 The 350-foot-tall Latting Observatory. In 1853, this was the highest manmade structure in North America.

sider the example I gave of stuffing a large circle into a small sphere.) A whale's volume equivalent could be *contained* within the sphere, but to do so would require the whale be first put through a meat-grinder that produces pieces no larger than the diameter of the sphere. (It would help if the whale could be folded or crumpled in higher dimensions like a piece of paper.) This therefore answers question 3. Similarly, for question 4, you could fit an infinite number of 3-D marbles into the 16-D sphere mentioned. Finally, just as a circular plate in two dimensions has zero thickness—and hence no volume—the whale, the earth, the Latting Observatory, and Einstein's brain have no "hypervolume" in higher dimensions. (Please forgive me for giving so many similar examples. I could have made my point by using two or three questions rather than six, but I hope the repetition reinforced the concept.) For some interesting student exercises, see note 4.

Hypersphere Packing

Now that we've discussed hyperspheres in depth, let's consider how hyperspheres might pack together—like pool balls in a rack or oranges in a box.

On a plane, no more than four circles can be placed so that each circle touches all others, with every pair touching at a different point. Figure 4.25 shows two examples of four intersecting circles. In general, for n-space, the maximum number of mutually touching spheres is $n + 2$.

What is the largest number of spheres that can touch a single sphere (assume that each sphere has the same radius)? For circles, we know the answer is six (Fig. 4.26). For spheres, the largest number is twelve, but this fact was not proved until 1874. In other words, the largest number of unit spheres that can touch another unit sphere is twelve. For hyperspheres, it is not yet known if the number is twenty-four, twenty-five, or twenty-six, nor is a solution known for higher dimensions, as far as I know. Mathematicians do know that it is possible for at least 306 equal spheres to touch another equal sphere in nine dimensions, and 500 can touch another in ten dimensions. But mathematicians are not sure if more can be packed!

Fact File

For those of you with a fondness for numbers, I close this section with a potpourri of fascinating facts.

- A cube has diagonals of two different lengths: the shorter one lying on the square faces and the longer one passing through the center of the cube. The length of the longest diagonal of an n-cube of side length m is

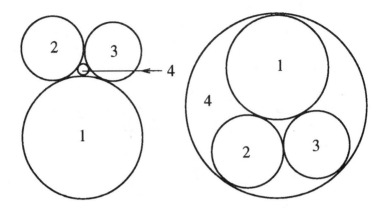

Figure 4.25 Circle packing. (left) In two dimensions, no more than four circles can be placed so that each circle touches all the others, with every pair touching at a different point. What happens in higher dimensions? (right) An attractive computer-graphic study of circle packing.

$m \sqrt{n}$. This means that if I were to hand you a three-foot-long thigh bone and ask you to stuff it into a 9-D hypercube with edges one foot in length, the bone would just fit because $\sqrt{9} = 3$. A dinosaur bone ten feet long could fit diagonally in a 100-D cube with edges only foot in length. A mile-long toothpick could fit inside an n-cube with edges the same length as those of an ordinary sugar cube, if n is large! On the other hand, a hypersphere behaves somewhat differently. An n-sphere can never contain a toothpick longer than twice its radius, no matter how large n becomes. As we've discussed, other odd things happen to spheres as the dimension increases.

- The number of edges of a cube of dimension n is $n \times 2^{n-1}$. For example, the number of corners of a 7-D cube is $2^7 = 128$, and the number of edges is $7 \times 2^6 = 7 \times 64 = 448$. Another factoid: two perpendicular planes in four-space can meet at a point.

- A 4-D analog of a pyramid has a hypervolume one-fourth the volume of its 3-D base multiplied by its height in the fourth direction. An n-dimensional analog of a pyramid has a hypervolume $1/n$ times the volume of its $(n-1)$-dimensional base multiplied by its height in the nth direction.

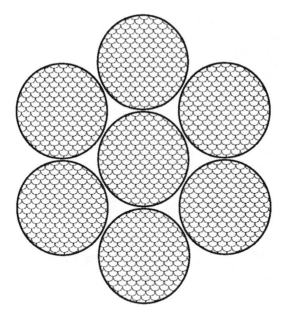

Figure 4.26   In two dimensions, a circle can make contact with six other circles of the same size. What happens in higher dimensions?

Scully: Just because I can't explain what I saw, doesn't mean I'm going to believe they were UFOs.

Mulder: Unidentified Flying Objects. I think that fits the description pretty well. Tell me I'm crazy.

Scully: You're crazy.

—"Deep Throat," *The X-Files*

five

He watched her for a long time and she knew that he was watching her and he knew that she knew he was watching her, and he knew that she knew that he knew; in a kind of regression of images that you get when two mirrors face each other and the images go on and on and on in some kind of infinity.

—Robert Pirsig, *Lila*

It is as if a large diamond were to be found inside each person. Picture a diamond a foot long. The diamond has a thousand facets, but the facets are covered with dirt and tar. It is the job of the soul to clean each facet until the surface is brilliant and can reflect a rainbow of colors.

—Brian Weiss, M.D., *Many Lives, Many Masters*

There are two ways of spreading light: to be the candle or the mirror that reflects it.

—Edith Wharton, *Vesalius in Zante*

I photocopied a mirror. Now I have an extra photocopy machine.

—Anonymous Interneter

mirror worlds

FBI Headquarters, Washington, D.C., 9:00 P.M.

You look out the window. The rain has stopped. "Let's continue our discussion in the fresh air," you say to Sally. "I'm in the mood for river sights." You carry a bag with props that will be useful later in your lessons.

Third Street winds downhill—past Leo's Pizza and Pasta, past life-size holograms of Clinton, Reagan, and Carter—following the kinks and bends of the Potomac River. The buildings are set well back from the streets and it isn't until you get nearer to the water that you see a considerable number of people. Hundreds. The nearby avenues are themselves very busy with endless streams of noisy taxis. A few joggers pass—no matter how swiftly they run, they are constrained to the asphalt and concrete and probably always would be. Would they want to experience, as you do, a jog into the fourth dimension?

You gaze into a clothing shop and stare at a large mirror. "Sally, I just love mirrors. Do you?"

Sally stares at you in the mirror. "Can't say that I thought much about it."

"Did you know that present-day mirrors are made by spraying a thin layer of molten aluminum or silver onto glass in a vacuum?"

"That's very interesting, I'm sure."

You duck into an alleyway and withdraw a small object from your bag. "Look at this Sally! I have a rotation machine."

Sally spins round and round, like a top. "Anyone can rotate."

"No, this device will let us rotate in the fourth dimension so that we are transformed into our mirror image!"

Sally smiles. "You've got to be kidding!"

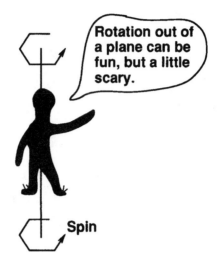

Figure 5.1   Rotating a 2-D person about a central axis.

"Not at all. Imagine a 2-D man living in Flatland. Pretend his right eye is brown and his left eye is blue. He wakes up one day and his wife screams, because his eyes have switched places. What actually happened is that a 3-D being rotated him about the center of his body into the third dimension" (Fig. 5.1).

You hand Sally a card with the words:

IN A SPACE OF N DIMENSIONS
THE MIRROR IS A SURFACE OF $N - 1$ DIMENSIONS

"Sally, in one dimension, objects are mirrored (reflected) by a point. In two dimensions, figures are mirrored by a line. In our world, mirrors are planes. In four dimensions, mirrors are solid!" You pause. "If a Flatlander could be rotated about a line in his plane, he would be turned into his mirror image. In general, if an object could be rotated around a mirror by moving into the higher dimension, the object would become its mirror image."

"What have you got in your hand?"

You hold up your left hand. "Nothing."

"Silly, your other hand."

"I've designed a machine that will rotate me into the fourth dimension. I haven't figured out a way to entirely vault into the fourth dimension, but even if I have to keep a piece of me here all the time, I can do some marvelous tricks. But don't worry, the machine is perfectly safe."

You place the device on the cement sidewalk and step on it. "This will turn me over in the fourth dimension so that I will be rotated into a mirror image of myself. The machine rotates me about a plane that cuts through my body from head to groin. Watch."

Sally stuffs her fist into her mouth and gasps.

"I am in complete control. Watch as my right half moves delta and my left half moves upsilon. When a Flatlander rotates, all you see is a line segment until his entire body fills the plane again. Similarly, you are now seeing a planar cross section of my body with all of my internal organs."

"Stop it!" Sally says as she looks at wriggling splotches of crimson floating in the air. Occasionally, the white of bone appears. It's as if a torturer from the Spanish Inquisition had sliced you with a huge sharp razor from head to toe and proudly brandished the thin slice in front of Sally (Fig. 5.2).

"Sally, as I rotate, all that remains is my cross section. Looks like sliced meat. An Omegamorph could turn us into our own mirror images by rotating us, in the fourth dimension, around planes that cut through our bodies. It's just like the Flatlander rotated about a line into the third dimension and then back down into the second dimension."

Sally steps back. "This can't be safe. Rotate yourself back to normal this instant."

"Sally, if you wanted to, you could move in close and tickle my kidneys."

"You *are* sick."

After a minute, you rotate back into the 3-D universe. You have returned to normal, except that you are your mirror image. "?kool I od woH"

"What did you say?"

You stare uncertainly at Sally.

Sally's eyes glisten like diamonds. For an instant you feel like light trapped within a shiny sphere. You peer at her corneas, your image many times reflected as if you are standing on the periphery of some gigantic crystal, alone in a field of darkness.

One of her earrings has fallen into a puddle. The puddle shivers. You are mesmerized by the lovely play of light on the clear rippling surface.

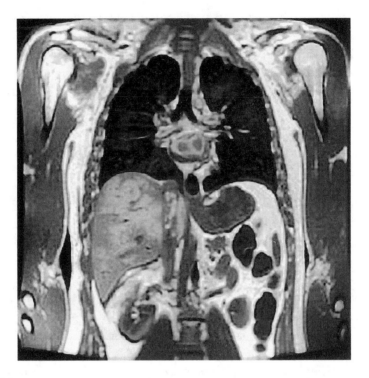

Figure 5.2 Rotating a 3-D person about a central plane. While the person is in the process of rotating, all that remains in our space is a cross section resembling microtomed meat.

You need not move. There is enough motion from the lights. You feel like you can live forever, suspended in space.

The light is reflected according to mathematical laws. Angles, polarizations, intensities, refractions, diffractions, interferences, geometrical optics, spherical aberrations. Such beauty from pure math.

Sally snaps you from your reverie. "My God, when you rotated, you looked like a CAT scan or an MRI from a hospital. The fourth dimension would make the ultimate medical diagnostic tool." She pauses. "But now you look a little different."

"Yes, my left and right sides are reversed. I couldn't do this without rotating in the fourth dimension."

"Can we prove it?"

"Listen to my heart. There's no way my heart could be on my right side without a rotation in the fourth dimension."

Sally puts her head on the right side of you chest. "I hear a strong beat. Holy mackerel, you've got to be correct. But aren't you going to change yourself back?"

"No, I want to see what it's like to spend a day as my mirror image. Imagine the romantic possibilities. Perhaps it alters one's perceptions. You are quite beautiful now."

Sally snaps her fingers and beckons you to follow. "You sound drunk, if you don't mind my saying so. Let's change the subject."

Sally's face begins to fade in the bright light. Even on your informal outings, she wears a stiff-collared suit and sober tie. One of the hardest things to take about her is the way her slender fingers dance when they are nervous.

"Sally, sorry. It must be the pressure of working long hours." You pause. "Do you know much about the occult?"

"Nothing, except that you are obsessed with it."

You nod. "A lot of so-called occult phenomena could be explained by the fourth dimension. Throughout history, some people have believed that spirits of the dead are nearby and can contact us. Of course, I can't believe this without proof, but lots of people believe that spirits can make noises, move objects, send messages—and there have been scientists who have used 4-D theory in an attempt to prove the existence of spirits and ghosts. The idea of 4-D beings just a few feet upsilon or delta from us had great popularity in the nineteenth century. In the seventeenth century, Cambridge Platonist Henry More suggested that a person's soul is physically unobservable because it corresponds to some hyperthickness in a 4-D direction. A dead person loses this hyperthickness. Henry More didn't use the term 'fourth dimension' but he meant just that."

"You said nineteenth century. What happened then?"

"Johann Carl Friederich Zöllner promoted the idea of ghosts from the fourth dimension. He was an astronomy professor at the University of Leipzig and worked with the American medium Henry Slade. Zöllner gave Slade a string held together as a loop. The two loose ends were held together using some sealing wax. Amazingly, Slade seemed to be able to tie knots in the string. Of course, Slade probably cheated and undid the wax, but if he could tie knots in the sealed string, it would suggest the existence of a fourth dimension."

"What makes you say that?"

You hand Sally a string with a piece of wax sealing the two ends (Fig. 5.3). "A 4-D being could move a piece of the string upsilon out of our

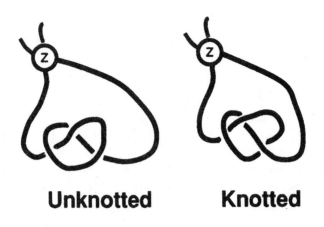

Unknotted   **Knotted**

Figure 5.3   A 4-D being would be the ultimate Houdini and could knot or unknot a string by temporarily lifting it into the fourth dimension. On the left is a string before it has been knotted. (Zöllner tried to have the left string transformed to the right without breaking the wax circle at top.)

space. This would be like cutting the string in the sense that the string could be moved through itself to form a knot. Once the string is in the correct orientation you can move it back delta into our space, and a knot would be tied without moving the ends of the string."

Sally traces a squiggle on some dirt using the tip of her shoe (Fig. 5.4). "I don't think a string can be knotted in 2-D space."

"You're right. In Flatland, there's no way a line can cross over itself. In fact, a string or a line can only be knotted in 3-D space. And any knot you tied in 3-D space will not stay tied in 4-D space because the additional degree of freedom will cause a knot to slip through itself." You pause. "By analogy, in 4-D space, a creature can knot a plane (surface), but this plane won't stay knotted in 5-D space. And the knotted plane cannot be formed in 3-D space."

Sally tugs your hand so that you start walking. "How in the world could you knot a plane?"

"Take a knotted line and then move it upsilon in the fourth dimension. The trail it traces will be a knotted plane. It never intersects itself. Of course, if we simply leave a trail in three-space as we move a knot, it will intersect itself, but since upsilon is perpendicular to all directions in our space, the 4-D knotted plane will not intersect itself."

Figure 5.4   A string cannot be knotted in 2-D space.

You start walking again. The city's fashionable shopping section has given way to a series of smaller shops. Many of them look a little seedy. You walk slowly, passing windows of secondhand clothing stores trying to pass themselves off as grunge boutiques. Some of the store signs read

SAVE JOBS—BUY PRODUCTS MADE IN THE THIRD DIMENSION.

One of the stores is called Earthlings Unite and it sells a startling array of goods—handcuffs, skimpy nightgowns, and golden watches displayed on black velvet. A woman in the store, perhaps the manager, has a gaudy punk hairdo—half purple, half orange—and looks as if she might weigh around eighty pounds. She stares at you and Sally and grins, revealing two large white canine teeth. They remind you of the teeth of some small but dangerous animal—a hyena, perhaps.

You turn back to Sally. "Zöllner devised three tests for Slade to see if Slade could use the fourth dimension to perform miracles. One: He gave Slade two oak rings that were to be interlocked without breaking them. Two: He gave Slade a snail shell and watched to see if a right spiral could be turned into a left spiral, and vice versa. Three: He gave Slade a rubber band and asked him to place a knot in one strand of the band. Actually it was band made from dried gut, but you get the idea."

"Could Slade do any of those things?'

"No. Nevertheless, the fourth dimension continued as an amusing interest for laypeople and scientists. A. T. Schofield, in his 1888 book *Another World*, suggested that

A higher world is not only possible, but probable; such a world may be considered as a world of four dimensions. Nothing prevents the spiritual world and its beings, and heaven and hell, being by our very side.

Sally sits down on a bench and begins to pull off her shoes. "Man, my feet are killing me."

You sit down next to her, and edge closer, and she stays where she is. Her eyelashes flutter, long and smoky against smooth, white cheeks. Here all is quiet, mostly deserted. You feel strangely alone, as if you and Sally can look down upon the world without being part of the world.

You turn to Sally. "It's good to be alone with you."

You feel her tremble slightly.

"I wish you wouldn't talk like that."

"Why?" you ask.

"I just prefer you didn't."

You feel as if you have pricked some layers of her psyche that she wishes to keep to herself. And yet you cannot stop. "You know, I sometimes feel you're not with me."

"I'm almost always with you."

And again you feel that she is being defensive.

"Will you stay with me if I keep talking? Sometimes I don't feel real. When I saw that Santa Claus man, I had the same feeling I get when I'm with you, a kind of fantasy."

"Tell me about him."

"I can't say much. His tesseract tattoo is odd. I know it's crazy to think he's following me, but I guess anything's possible in Washington."

"True. It's hard to separate illusion and reality. Sometimes I don't even try."

"Was I imagining the Santa Claus man? Am I imagining you?"

"If you're imagining me, what of it? Do you enjoy it?"

"Yes."

"Then that's enough."

"I wish it were, but I'm the kind of person who wants answers to everything."

"Me too." She places her hand on yours. "But you are sometimes too intense. It scares me. You need to get more rest. It's late. Let's finish up our discussion on the fourth dimension."

You nod. "Sally, a single 4-D God might appear as many Gods or angels as He intersects the 3-D world. We could all be part of some 4-D entity."

"I don't buy it. How could we all be part of a 4-D creature?"

"Sally, remember Figure 2.18 in which the footsteps of a 4-D creature appear as two separate humans?"

"What are you talking about? 'Figure 2.18?' We're not in some kind of book."

You nod. "Here's what P. D. Ouspensky said about it in his 1908 essay 'The Fourth Dimension':

We may have very good reason for saying that we are ourselves beings of four dimensions and we are turned towards the third dimension with only one of our sides, i.e., with only a small part of our being. Only this part of us lives in three dimensions, and we are conscious only of this part as our body. The greater part of our being lives in the fourth dimension, but we are unconscious of this greater part of ourselves. Or it would be still more true to say that we live in a four-dimensional world, but are conscious of ourselves only in a three-dimensional world.

In another half hour, you are walking Sally back to her car at the FBI headquarters parking garage. The hard heels of her black shoes resound hollowly in the large, cement structure.

"Sally, tomorrow, we try to save the president."

"I'll come to your office at eight."

Once back in your office, you hold a handkerchief against your mouth and nose. The smell of the place is much worse than you remember it. There is a heavy smell, an odor of putrefaction. Probably the smell of some of your leftovers or the remnants of some McDonald's hamburger in the garbage.

You remove a sleeping bag from under your desk. For a moment, you leave open the door to your office, to listen to any sounds in the hallway. An FBI agent is often alone. Maybe you should have married when you had the chance, years ago. But the timing was not right.

Would it be this way for the rest of your life? Are you always to live mostly in your office, smile politely at the custodians, do your work, and eat sandwiches from Leo's Deli by yourself?

You turn off the lights and stare out the window. Outside there is a drizzle, and lamps throw broken yellow gleams off puddles. You are hungry, and find two apples to chew on. It is so dark they remind you of hyperspheres. Somehow the apples don't taste right when you can hardly see them.

You sit in the darkened office with only the lights from the streets reflecting off your shiny desk surface. You feel as if you are in a dream, sitting in limbo. Is Sally right—is your dream of lifting off into the fourth dimension unrealistic? Does it matter? What would it prove if you did? You know you can go on for months, perhaps years, trying to unravel the fourth dimension without ever fully understanding it.

At first you think it might be a good idea to try an experimental flight to prove your theories. A flight into the fourth dimension is quite a risk. How could you dare? Yet some inner compulsion drives you. You need not "experiment." You decide to take the plunge. Tomorrow would be the day.

You press a button and Beethoven music pours into the room like water, water you have looked into, water you have held.

The Science Behind the Science Fiction

The curious inversion of Plattner's right and left sides is proof that he has moved out of our space into what is called the Fourth Dimension, and that he has returned again to our world.

—H. G. Wells, "The Plattner Story"

That was why Mick had looked funny; he had turned over in hyperspace and come back as his mirror-image.

—Rudy Rucker, *Spacetime Doughnuts*

"Could I but rotate my arm out of the limits set to it," one of the Utopians had said to him, "I could thrust it into a thousand dimensions."

—H. G. Wells, *Men Like Gods*

Zöllner Experiments

Early widespread interest in the fourth dimension did not take place in the scientific and mathematical communities, but among the spiritualists. The American medium and magician Henry Slade, became famous when he was expelled from England for fraud connected with spirit writing on slates. Astronomer J. C. F. Zöllner was almost completely discredited because of his association with spiritualism. However, he was correct to suggest that anyone with access to higher dimensions would be able to perform feats impossible for creatures constrained to a 3-D world. He suggested several experiments that would demonstrate his hypothesis—for example, linking solid rings without first cutting them apart, or removing objects from secured boxes. If Slade could interconnect two separate unbroken wooden rings, Zöllner believed it would "represent a miracle, that is, a phenomenon which our conceptions heretofore of physical and organic processes would be absolutely incompetent to explain." Similar experiments were tried in reversing snail shells and tying knots in a closed loop of rope made of animal gut. Perhaps the hardest test to pass involved reversing the molecular structure of dextrotartaric acid so that it would rotate a plane of polarized light left instead of right. Although Slade never quite performed the stated tasks, he always managed to come up with sufficiently similar evidence to convince Zöllner and these experiences became the primary basis of Zöllner's *Transcendental Physics*. This work, and the claims of other spiritualists, actually had some scientific value because they touched off a lively debate within the British scientific community.

IS THE UNIVERSE SHAPED LIKE A PRINGLES POTATO CHIP?

Hyperthickness

Is it possible that our space has a slight 4-D hyperthickness? If every object in our space is a millimeter thick in the direction of the fourth dimension, would we notice this 4-D component of our bodies? If we are actually 4-D creatures, and our bodies are only 3-D cross sections of our full bodies, how would we know?

As I've mentioned repeatedly, the possibility of a fourth dimension led to religious debate over the centuries. Spiritualists have even wondered whether the souls of our dead drifted into another dimension. For example, British philosopher Henry More argued in *Enchiridion Metaphysicum* (1671) that a nether realm beyond our tangible senses was a home for ghosts and spirits. His descriptions were not too far from how modern mathematicians describe a fourth dimension. Nineteenth-century theologians, always searching for the location of Heaven and Hell, wondered if they could be found in a higher dimension. Some theologians represented the universe as three parallel spaces: the Earth, Heaven, and Hell. Theologian Arthur Willink believed that God was outside of these three spaces and lived in infinite-dimensional space. Karl Heim's theology, described in his book *Christian Faith and Natural Science*, emphasizes the role of higher dimensions. Several philosophers have suggested that our bodies are simply 3-D cross sections of our higher 4-D selves.

Aliens with Enantiomorphic Ears

Although the vague notion of a fourth dimension had occurred to mathematicians since the time of Kant, most mathematicians dropped the idea as fanciful speculation with no possible value. They had not discussed the fact that an asymmetric solid object could, in theory, be reversed by rotating it through a higher space. It was not until 1827 that August Ferdinand Möbius, a German astronomer, showed how this could be done—eighty years after Kant's papers on dimension.

If you encountered a Flatlander, you could, in principle, lift the Flatlander out of his plane and flip him around. As a result, his internal organs would be reversed. For example, a heart on the left side would now be on the right. Similarly, a 4-D being might flip us around and reverse our organs. Although such powers are, in principle, possible within the auspices of hyperspace physics, I should remind readers that the technology to manipulate space in this fashion is not possible; perhaps in a few centuries we will explore hyperspace in ways today only dreamed about in science fiction.

Many creatures in our world, including ourselves, are bilaterally symmetric; that is, their left and right sides are similar (Fig. 5.5). For example, on each side of our bilaterally symmetric body is an eye, ear, nostril, nipple, leg, and arm. Beneath the skin, our guts do not exhibit this remarkable symmetry. The heart occupies the left side of the chest; the liver resides on the right. The right lung has more lobes than the left. Biologists trying to explain the origins of left-right asymmetries have recently discovered several genes that prefer to act in just one side of a developing embryo. Without these genes, the internal organs and blood vessels go awry in usually fatal ways. Mutations in these genes help explain the occurrences of children born with their internal organs inverted along the left-right axis, a birth defect that generates remarkably few medical problems. It is imaginative to consider this as a "disease" of the fourth dimension. If we had 4-D powers, we might be able to reverse some of these strange asymmetries.

One way to visualize the flipping of objects in higher space is to consider the two triangles in Figure 5.6. These are called "scalene" triangles because they have three different side lengths. They make an "enantiomorphic pair" because they are congruent but not superimposable without lifting one out of the plane. Similarly, in our 3-D world, there are many examples of enantiomorphic pairs—these consist of asymmetric solid figures such as your right and left hands. (If you place them together, palm to palm, you will see each is a mirror reflection of the other.) The scalene triangles, like your two hands, cannot be superimposed, no matter how you rotate and slide them. However, by rotating the triangles around a line in space, we can superimpose one triangle on its reflected image. Similarly, your own body could be changed into its mirror image by rotating it around a plane in four-space. (See Appendix B for information on Wells's "The Plattner Story" and the adventures of a chemistry teacher whose body is rotated in the fourth dimension.)

In four dimensions, figures are mirrored by a solid. Mirrors are always one dimension less than the space in which they operate.

If there were a hyperperson in four-space looking at our right and left hands, to him they would be superimposable because he could conceive of rotating them in the fourth dimension. The same would apply to seashells with clockwise and counterclockwise spirals as in Figure 5.7.

Can you think of other examples of enantiomorphic pairs in our universe? For example, your ears are enantiomorphs. (I like to imagine a race of aliens

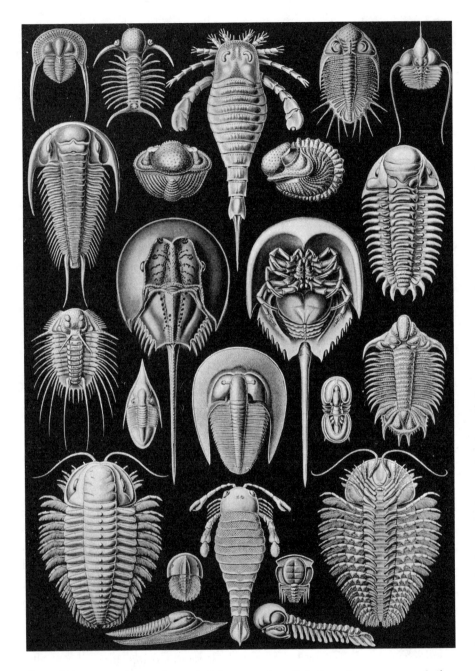

Figure 5.5 Horseshoe crabs and various species of their extinct ancestors (trilobites), all exhibiting bilateral symmetry.

Rotate

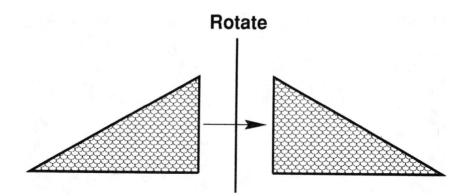

Figure 5.6 The two triangles can be superimposed only if one is first rotated up out of the page into a higher dimension.

whose right and left ears are identical, not enantiomorphic. Can you imagine what they might look like?)

Möbius Worlds

If our entire universe were suddenly changed into its mirror image, would we perceive a difference? To answer this question, consider a Lineland on which reside only three slimy aliens: "Thing 1," "Thing 2," and "Thing 3," all facing east—that is, they are all looking to the right.

_____Thing 1 ___ Thing 2 ___ Thing 3 _____Lineland

If we reverse Thing 2, the change will be apparent to Thing 1 and Thing 2. But if we reverse the entire line of Lineland, the 1-D aliens would not perceive a change. We higher-dimensional beings would notice that Lineland had reversed, but that is because we can see Lineland in relation to a world outside it. Only when a *portion* of their world has reversed can they become aware of a change. The same would be true of our world. In a way, it would be meaningless to say our entire universe was reversed because there would be no way we could detect such a change. Why is our world a particular way? Philosopher and mathematicians Gottfried Wilhelm Leibniz (1646–1716) believed that to ask why God made the universe this way and not another is to ask "a quite inadmissible question."

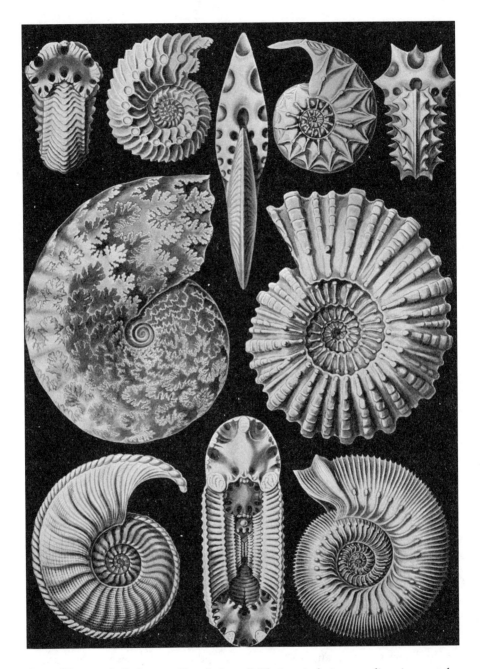

Figure 5.7 Seashells have a "handedness." Their spirals in one direction can be transformed into spirals in the other direction by rotating the shells through the fourth dimension. (Shell growth is usually not confined to a plane but also extends in a third direction, like the swirl of an ice cream cone.)

To get a better understanding of Leibniz's comment, consider a 2-D Flatland. To mirror reverse the entire Flatland universe, all we have to do is turn the plane over and view it from the other side. In fact, we don't have to turn the world over. Consider Flatland to be like a vertical ant farm in which the ants are essentially confined to a 2-D world. The world is a left-handed world when viewed from one side of the glass and a right-handed world when viewed from the other. In other words, Flatland does not have to change in any way when you view it from one side or the other. The only change is in the spatial relation between Flatland and an observer in three-space. In the same way, a hyperbeing could change his position from upsilon to delta and see a seashell with a right-handed spiral become a left-handed spiral. If he could pick up the shell and turn it over it, would be a miracle to us. What we would see is the shell disappear and then reappear as its mirror image. This means that enantiomorphic structures are seen as identical and superimposable by beings in the next higher dimension. Perhaps only a God existing in infinite dimensions would be able to see all pairs of enantiomorphic objects as identical and superimposable in all spaces.

There are other ways of turning you into your mirror image, without your ever leaving the space within which you live. Consider a Möbius strip, invented in the mid 1800s by August Möbius. A Möbius strip is created by twisting a strip of paper 180 degrees and then taping the ends together. (A Möbius strip has only one side. If that's hard to believe, build one and try to color one side red and the "other" side green.) By way of analogy, if a Flatlander lived in a Möbius world, he could be flipped easily by moving him along his universe without ever taking him out of the plane of his existence. If a Flatlander travels completely around the Möbius strip, and returns, he will find that all his organs are reversed (Fig. 5.8). A second trip around the Möbius cosmos would straighten him out again.

A Möbius band is an example of a "nonorientable space." This means, in theory, it is not possible to distinguish an object on the surface from its reflected image. The surface is considered nonorientable if it has a path that reverses the orientation of creatures living on the surface, as described in the previous paragraph. On the other hand, if a space preserves the handedness of an asymmetric structure, regardless of how the structure is moved about, the space is called "orientable."

Just as on the Möbius strip, strange things would happen if we lived on the surface of a small hypersphere. By analogy, consider a Flatlander living in a universe that is the surface of a small sphere. If the Flatlander travels along the

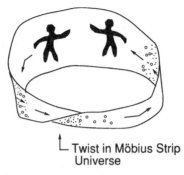

└─ Twist in Möbius Strip
Universe

Figure 5.8 A 2-D human on a Möbius strip universe. If the human travels around the strip, his internal organs will be reversed.

sphere, he returns to his starting point. If he looks ahead, he sees his own back. If you lived in a hyperspherical universe, you too could return to your starting point after a long time. If the hypersphere were small, you'd see your own back while looking forward. As alluded to in the section on extrinsic geometry (Chapter 1), some cosmologists have suggested that our universe is actually a large hypersphere. The universe may be finite but with no boundary, just as a sphere's surface is finite, but has no edge. In other words, our universe may be a 4-D sphere with a 3-D surface having a circumference of the order of 1000 billion light-years. (One light-year is the distance traveled by light in one year—about 5,900,000,000,000 miles.) According to this model, what we perceive as straight, parallel lines may be large circles intersecting at two points fifty billion light-years away in each direction on the hypersphere (in the same way that longitude lines on a globe actually meet at the poles.)

If our universe is curved, our space can be finite and still have no edge. It simply curves back on itself. This means that if we fly far through space, we could never encounter a wall that indicates that space goes no further. There could be no sign that reads:

GO BACK. SPACE ENDS HERE.

The idea that our universe could be the surface of a hypersphere was suggested by Einstein and has startling implications.[1] As an analogy, again consider a 2-D Flatland on the surface of a sphere. If an inhabitant started to paint Flatland's surface outward in ever-widening circles, he would reach the halfway point when the circles would begin to shrink with the Flatlander on the *inside*; eventually he would paint himself into a little place of the universe, at which

Figure 5.9 Computer graphic representation of a Klein bottle. (Computer rendition by the author; see Appendix F for program code.)

point he could paint no further. If the paint were toxic, a mad Flatlander using this approach could ensure that all life-forms were destroyed. Similarly, in Einstein's model of the universe, if a human began to map the universe in ever-expanding spheres, he would eventually map himself into a tiny globular space on the opposite side of the hypersphere.

Our universe could have other equally strange topologies like hyperMöbius strips and hyperdoughnuts, with additional interesting features that are beyond the scope of this book.[2] For example, in 4-D space, various surfaces containing Möbius bands can be built that have no boundary, just like the surface of a sphere has no boundary. The boundary of a disc can be attached to the boundary of a Möbius band to form a "real projective plane." Two Möbius bands can be attached along their common boundary to form a nonorientable surface called a Klein bottle, named after its discoverer Felix Klein (Fig. 5.9). The Möbius band has boundaries—the band's edges that don't get taped together. On the other

Figure 5.10   Glass Klein bottle designed and manufactured by glassblower Alan Bennett.

hand, a Klein bottle is a one-sided surface without edges. Unlike an ordinary bottle, the "neck" is bent around, passing through the bottle's surface and joining the main bottle from the inside. One way to build a physical model of a Klein bottle in our 3-D universe is to have it meet itself in a small, circular curve.

Imagine your frustration (or perhaps delight) if you tried to paint just the outside of a Klein bottle. You start on the bulbous "outside" and work your way down the slim neck. The real 4-D object does not self-intersect, allowing you to continue to follow the neck that is now "inside" the bottle. As the neck opens up to rejoin the bulbous surface, you find you are now painting inside the bulb.

If an asymmetric Flatlander lived in a Klein bottle's surface, he could make a trip around his universe and return in a form reversed from his surroundings. Note that all one-sided surfaces are nonorientable. Figure 5.10 is a glass Klein bottle created by glassblower Alan Bennett (see note 2 for more information). Figure 5.11 is a more intricate Klein-bottle-like object.

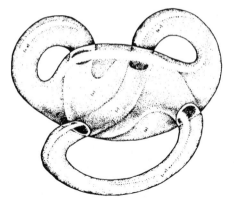

Figure 5.11 A more intricate model of a higher-dimensional object. [For more details on this shape, see Ryan, P. (1991) The earthscore notational system for orchestrating perceptual consensus about the natural world. *Leonardo.* 24(4): 457–65. The drawing was done by Gary Allen. The bottle object is called a relational circuit.]

Many kinds of spatial distortions are discussed in detail in my book *Black Holes: A Traveler's Guide.* We still need to learn more about the large-scale structure of our universe before we can determine whether orientation-reversing paths exist. Imagine the possibilities if these paths exist. When you traveled in a rocket ship and returned as your mirror image, all your screws, scissors, fonts, body organs, and clocks would have changed their orientations relative to your friends who never risked the journey. If your spouse or loved one returned to you reversed, would your feelings for them change? Would you notice the difference? Could they still drive your car, write in a manner legible to you, use a computer keyboard, digest the same foods, or read your books? Would the enantiomorphic molecules in their bodies also be reversed? Would there be any advantages to being mirror reversed? Would future societies, seeking uniformity, send out left-handed people in spaceships so that on their return they would be right-handed? Would governments purposefully send out people to be mirror reversed, thereby creating whole new segments of the population that could not mate with "normal" people or contract deadly pathogens that evolved to prey on biomolecules with particular enantiomorphic characteristics?

> *Scully:* Oh God, Mulder, it smells like . . . I think it's bile.
> *Mulder:* How can I get it off my fingers without betraying my cool exterior?
>
> —"Squeeze," *The X-Files*

six

The new Chief Circle encourages the Masses to worship Higher Space Beings as Angels and Gods. Higher Space is the royal road to that which is beyond all imagining. He that hath ears, let him hear.

—Rudy Rucker, *The Fourth Dimension*

For Catherine, time had lost its circadian rhythm; she had fallen into a tesseract of time, and day and night blended into one.

—Sidney Sheldon, *The Other Side of Midnight*

The Old Testament says that we cannot see God and live. It follows that there must be some second space, just as all-present as the three-dimensional space but completely unobservable.

—Karl Heim, *Christian Faith and Natural Science*

the gods of hyperspace

The White House, Washington, D.C., 9:00 A.M.

You are back in the White House at the exact location the president was last seen alive. A large Persian carpet decorates the floor; along the wall are potted plants and a few chairs. Nothing has changed, except for a large bronze replica of a Liberty Bell someone has moved to a corner of the room for refurbishing. Perhaps Vice President Chelsea Clinton is readying the bell as a gift to Sadaam Hussein who recently renewed his commitment to world peace.

There are sounds from all around the room, like the swooshing of brooms on wet cement. An ammonia odor fills the air.

"I suppose you have a plan to rescue the president?" Sally says as her eyes dart all around the room.

Before you can reply, there is a flash of light; in front of you is a moving shape. It floats right and left at a height of five feet. Then is stops and shoots downward. It is about the size of a pillow and covered with skin.

Sally withdraws her gun.

You jump toward her. "Don't shoot."

You slowly walk to the blob and touch it with your index finger. The odd thing shrinks to the size of a cantaloupe yet retains its skin-like texture. "It feels warm and soft, just like human flesh."

Now the blob increases in size again while you and Sally stare with your mouths agape. When it stops growing, it is the size of a large turkey. It is mostly flesh, but along one side is a ridge-like structure and a strip of black, velvet-like material.

The creature starts bouncing on the floor, making loud slapping sounds like the flapping of a fish out of water. Then it is quiet and still, resembling

141

a biological oddity at some museum of pathology. You recall seeing something like it at the Smithsonian Museum of Pathology where curators had unveiled freakish things afloat in jars, body organs delicately detached, soft arteries infused with wax, flesh still fresh as if the objects were still alive.

There are footsteps behind you. "What's going on here?"

A man approaches you with his hand on his gun. He stands six feet tall, with a barrel chest and rock-hard belly to match. His cheeks look as if they'd been sandblasted. The arms of his camouflage suit are torn off, revealing muscular arms decorated with faded green tattoos that run from his wrists to his biceps. His huge hands are scarred and soaked with sweat. With slow, deliberate movements, the man removes a black headband from his pocket and ties it around his forehead.

"FBI," you say and flash him your badge. "Who are you?"

"Captain Richard Narcinko, ex-Navy SEAL. My job is to secure the room. This is a matter of national security."

"Ooh," Sally says as a tiny dot of flesh rises from the blob. It drifts toward her, moving like a feather floating in the breeze. Sally's hand goes to her face as her eyes open wide in wonder. The dot floats at the height of her head, flashing various shades of red and lavender. At the edges of the fleshy ball are little sparkles, as if the dot is igniting tiny dust particles in the air as it moves through them.

Sally raises her gun again. "Oh my God!" she says as she points to another flesh-ball rising from the blob. It seems to follow the first.

But then she lowers her gun and smiles as the twin balls of flesh dance in front of her eyes. She doesn't seem frightened at all.

"Be careful," you say to Sally. "Back away."

"I don't sense they are harmful."

" 'Sense?' Sally, what are you talking about? Back away."

Outside there is an occasional cry of a bird. These cries sound distant, diffuse, as if part of a reality isolated from what is happening in the White House.

One of the dots continues to dance before Sally's eyes. The other slips between her legs, rises along her back, and merges with the first dot. Sally's grin widens, and her whole body shivers like an excited child's. A few strands of her hair stand at right angles to her body; perhaps there is static electricity in the air. She raises her hands as if she is conducting an invisible orchestra, but it begins to look more like she is casting a spell or stirring a witch's potion.

Sally turns her head upward. "I feel—I feel a current running through my body. But it doesn't hurt."

More of her hair strands stand out from her body, as she spreads her legs and trembles slightly. She starts taking great gasps of air.

Narcinko crouches and tries to track the dot with his Heckler and Koch 9-mm semiautomatic.

"Don't shoot!" you yell at Narcinko.

Sally's breath comes in short spurts now. She is shaking.

Narcinko takes a step forward. Perhaps he is going to try to swat the dots of flesh away from Sally, but, as he approaches, the dots separate and stop dancing. They look delightful, but you worry. If the tiny dots are hot and penetrate Sally's body, she would burn. The dots seem to alter her perceptions, and that could also mean danger, a danger not to her body but to her mind.

"Wait," you say as a flowery odor fills the air.

The dots vibrate, making oboe-like sounds and moans. Could this be what 4-D speech sounds like as it intersects your world?

"Don't stop," Sally whispers.

The dots continue to vibrate. Sally breathes in short shallow breaths, oblivious of who would hear or who would care. Then she starts to rock back and forth as her pupils dilate.

You can't stand it any longer. The sounds of the flesh dots are affecting you too, making you drunk with pleasure. You lean into Sally. Her hair has the sweet but musty smell of a garden in early autumn after a long, luscious rain.

"Stop," Narcinko yells to the lights as he claps his hands together with a loud bang.

Narcinko leaps toward Sally. "Enough of this," he says as he swings his huge hands at the dots. But they are too quick for him.

The dots dart through the closed, bulletproof window and shoot toward Pennsylvania Avenue. Sally is the first out of the White House to follow them. You and Narcinko are close behind.

You try to get your sense of direction because intermixed with the spinning dots of flesh are flares coming from other flesh blobs, flickering so swiftly that you cannot be sure where they are coming from. The shimmering blobs reveal momentarily eerie forms, things that look like eyes swimming and swiveling to stare at you with an incandescent glare, like the eyes of owls or deer when a beam from a car's headlight catches them by surprise.

The air seems dry and hot, as if the dots emitted some heat. An odor, not quite perfume but more animal-like, permeates the air and fastens to you.

Most of the flesh dots disappear while the two original dots float like tiny bright bubbles toward the cars and streetlights. They oscillate all the colors of the rainbow, hover, and then begin to move rapidly. Sally giggles nervously as she watches the dots bounce from car to car like luminous balls in a Ping-Pong game. They go faster and faster and then spiral soundlessly down into a manhole cover in the road.

You jump when a flock of crows on fire flap low over your head, their "scree-scree-scree" calls shocking you almost as much as the balls of living flesh. You turn your neck to follow the birds' flight and realize they are not truly on fire. It is just the light from the orange sun on their shiny backs. You become conscious of a chill wind on your arms, a chill that makes you shiver.

You face Pennsylvania Avenue again, placing your hand over your forehead, searching for any signs of the floating flesh.

"Come back!" Narcinko calls to you from the White House steps. "For God's sake, look at this."

You and Sally run back and hear a whooshing sound as the odd flowery odors turned to something more ominous. The smell of vomit on wet hay.

"This way," Narcinko roars in a deep baritone voice.

Inside is a large quiescent blob with pink lips. Narcinko withdraws a large hunting knife from his pocket.

Sally shakes her head. "What are you going to do with that?"

"Just an experiment."

The skin is thick and he has to slice twice to draw blood.

Suddenly the blob goes wild. It decreases in size until it looks like several floating testicles. Four lumps of flesh are on the left of Sally. Another three or four are to your right. They race toward you, as you duck, and then they swiftly encircle Sally. She shields her face with her arms.

Then she disappears with a popping sound. Her screams suddenly stop as you hear their echoes reverberating through the White House.

Narcinko comes closer. "What has it done to her?"

You rush toward the spot where Sally last stood, but there is no trace of her. You wave your hands back and forth through the air but feel nothing. "You did it to her!" You turn to Narcinko. "Why did you cut the creature?"

"What's that?" Narcinko says.

About five feet away lies an eggplant-like lump of flesh that you had not noticed. It sits on the colorful Persian carpet and does not move.

Narcinko stares at the fleshy lump and a shiver runs up his body. "What the hell is it? What's happening here?"

"We were here to help rescue the president. He was abducted by 4-D creatures."

"How do *you* know this? That's classified."

"Narcinko, can you find me a strong bag and lots of rope?"

He eyes you suspiciously. "We heard rumors about the 4-D creatures. We're not telling the public much or else they may panic. The White House is officially closed today."

"I think I can save Sally and the president. Get me a bag and rope."

Narcinko dashes away and returns with a large burlap bag and about six feet of rope. "I got this from the gardner's closet."

You take the bag. "Excellent."

You slowly make your way over to the fleshy blob on the Persian carpet. Stooping down, you place the bag next to the blob and very gently push the blob into the bag. Narcinko brings you the rope and you tie the bag tightly shut.

As you look into Narcinko's face, you see his mouth open and close spasmodically. Why? What could be wrong?

You turn, and see that the lump of flesh is now out of the bag. It must have escaped quickly without making a sound. Your eyes had been off the bag for only a second or two. The rope is still tied tightly around the opening.

Narcinko looks closer. "It's different."

You nod. There are three pieces of it now, like pink hot dogs. Narcinko unties the knotted rope around the bag. He slowly places his hand inside as if he is about to plunge into ice-cold water. "Nothing's in here."

"I should have expected this. I think the creature is stuck in our world and wants to leave. This is just a cross section of him, a solid cross section that is stuck in our 3-D world." You explain some of the same things you have been telling Sally. "If I stick my foot in the surface of a pond, 2-D inhabitants on the pond's surface would see only a circle or an ellipse depending on what angle I intersect their world. If a 2-D creature tried to trap me in his world by tying a rope around me, I could withdraw my foot—unless the rope is very tight or there is a larger part of my body on each side of the pond's surface."

"But what about the girl and the president? Where are they?"

You point to the several shapes that quickly coalesce into a single blob. "The struggles of this creature vaulted them into hyperspace."

You pace back and forth. How could only a piece of an Omegamorph get stuck in your world?

Somehow you must save Sally and the president. You are also worried about the creature escaping. You want it here to study. Perhaps you can get it to help you rescue Sally and the president in return for helping it get back into its own world. It is your only link to Sally.

How can you make sure it stays in the White House? It is a 3-D section of the true creature. If you tie the burlap bag extremely tight around the blob, or blobs, perhaps you could keep it in this world.

Again, you gently place the blob inside the bag, and this time you tie the bag tightly around the creature with all your might. Next you tie the other end of the rope to the large bronze replica of the Liberty Bell.

You turn to Narcinko. "I'm going into hyperspace to search for them. But I want something stronger than rope. Can you find me some chains? Also I need two flares and a photo of the president."

"What do you need all this for?"

"Just do it."

Narcinko looks like he is about to punch you, but then he walks away down a dimly lit hallway.

Fifteen minutes later, Narcinko returns with some chains made of a light, strong alloy. He hands you two flares that you stuff in your pocket along with a photo of the president.

Narcinko helps you tie the chain in a harness-like configuration around your shoulders and waist. The loose chain is coiled on the floor, three hundred feet of it, and the other end tied to the Liberty Bell.

Narcinko steps back suddenly when the thing in the bag starts writhing like a worm.

"I'm going to leap upward into hyperspace. I'll signal you if there's trouble by tugging on the chain once quickly followed by two long tugs. Pull me back if you feel that."

Narcinko looks from you to the pulsating bag then back to you. "This is crazy."

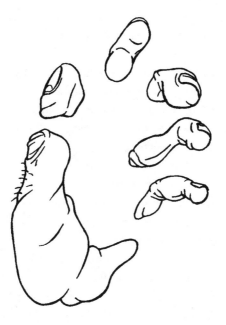

Figure 6.1 A portion of your hand disappears into another dimension. (Drawing by April Pedersen.)

"We've got to give it a try. If I can find just the right location in this room where it seems to be easy to leave three-space, I should be able to jump."

First you place your hand in the air and watch a piece of the palm seem to disappear (Fig. 6.1). Next, you leap two feet into the air and are pulled back down by the chain.

Perhaps it is a bit too heavy. You leap again and feel a queasy feeling in the pit of your stomach, as if someone has placed a hook there and is trying to pull your stomach up through your esophagus. You jump again, suddenly hear a hissing sound, and for an instant you feel weightless.

You look down and think you see large apartments honeycombed with light. In one apartment, you see right through the walls and observe a woman, nearly naked, watching television. In the distance are honking horns and the voices of thousands of people.

As you fly, you feel there is light and music coming from Washington, D.C. It is as if thousands of Washingtonians are celebrating, clapping their hands, cheering your upward flight. The headlights of tiny cars are wreathed in mist, making you feel as if you're rising up through a waterfall of stars.

Suddenly the sounds and lights disappear. You look down and are standing on something that looks like asphalt. You can't quite see the other end of the chain. You are somewhere in the fourth dimension!

You imagine that, to Narcinko, the chain simply ends in space, disappearing abruptly in midair. There is still tension on the chain, and as you pull it you can tell that more of the chain is coming toward you.

A vast expanse of space opens before you. Ecstasy! Surely if you were to cut the chain around your leg, you could fly even further. Physical laws and rules did not seem to be limitations, only stepping-off points for something greater, more profound.

Several feet away are some undulating pumpkins. Well, not pumpkins exactly, but close enough. As you walk you see glass, metal, and concrete structures—all in strange polyhedral shapes, like large, twinkling stars. A few greenish, frilly objects float in the air (Fig. 6.2). Plants?

You are drunk with pleasure, but realize that even though you are in a 4-D world, your 3-D perception is capturing only a piece of its true form. These must be the spatial cross sections of buildings, curbs, streetlights, bushes, and such. Some of the beings resemble blobs of flesh. Some are covered with fabric; some are totally naked. You reason that you must be in a world parallel to your own, elevated upsilon in the fourth dimension.

What would a Flatlander see as he floated in your world? If he could only *perceive* in two dimensions, he could only see a piece at a time. Imagine his confusion as he only saw cross sections of objects: desks, lights, couches, tables, plants. . . . It might drive him mad.

But you are not mad. You breathe more easily now; your heart is beating slower than a few seconds before. A cool, gentle breeze dries the perspiration on your forehead. As you look in one direction, you see a city-like structure with iridescent windows that make the fog radiate and pulse with life of its own.

Suddenly, three blobs of flesh coalesce in front of you. Leather begins to surround one of them. Leather? Mad visions of 4-D cows float through your brain.

Then the blobs disappear. Perhaps they are citizens of hyperspace who had not noticed you, or had seen you but had not cared. Would they be used to penetration of 3-D worlds into their own?

Figure 6.2   The cross section of 4-D plants?

To prove you have some control over your destiny, despite the chain around your leg, you swing away from the path you've taken and follow a little green light that sits atop a metallic structure. You pass tiny craters and odd blobs of maroon that look neither alive nor dead, but as if they are hibernating. Ah, if only you had time to take a more leisurely travel in this world. But you have responsibilities.

How can you ever find Sally in such a strange world? She could be inches away from you and you wouldn't be able to see her if she were not in your "plane" of vision. You can't see beyond your three dimensions.

"Sally," you scream. There is no response, except perhaps a barely audible laughter. You trod through fields of strange vegetation surrounded by puffs of mist. It seems you are floating from one spatial "plane" to another. Then the chain becomes tight. You can go no further.

You bring out your flare, light it, and wave it around. You scream and scream.

Eventually a small ball of flesh comes near and you wave the flare even faster. The ball jumps up and down and grows larger. When the cross section increases to the size of a large pumpkin, you rub it gently and it vibrates in response. There is a purring sound. You pull the creature delta. A dozen ellipsoidal blobs coated with a hard enamel-like surface come into view (teeth?) and finally you see a perfectly white ball with a trace of blood vessels. You hope this white ball is the cornea of the creature's hyperspherical eye. After a few seconds, it transforms into a brown sphere with musculature like the iris of an eye, a perfect ball. This ball is so shiny that it sparkles in the light. At last you see what you've been waiting for: two, black, moist orbs, the size of baseballs, floating in front of you. These must be the creature's pupils.

You feel a strange shiver go up your spine as you look into the Omegamorph's glistening eyes. You feel a chill, an ambiguity, a creeping despair. The Omegamorph is still. Neither of you moves. Its eyes are bright. If you could see it smile, you imagine it would be relentless and practiced. Time seems to stop. For a moment, your mind fills with a cascade of mathematical symbols. But when you shake your head, the formulas are gone. Just a fragment from a hallucination. But the inscrutable Omegamorph remains.

"Thank you for paying attention to me," you say. You bring out a photo of the president and one of Sally that you've had in your wallet for years. "Please, bring my friends back and push them into my world. As

Figure 6.3   If Lucifer were a 4-D being in Heaven, five feet upsilon from our 3-D "plane," and then expelled delta below our world, could he, for a time, have gotten stuck in our universe? This amusing fantasy leads to a good question: If there were a fourth dimension, how come we seem to have such difficulty accessing it? (Drawing by Brian Mansfield.)

payback, I'll guide you to the creature we have in the bag. And then we will release the creature."

Can the creature understand you? The two black spheres just stare at both you and the photos and then, suddenly, hundreds of moist, baseball-sized flesh-balls come near you and you feel as if you're being pushed down. There is a deep roaring sound. Is this what Lucifer felt as he was expelled from Heaven? (Fig. 6.3).

You tumble to the White House floor along with Sally and the president.

There is a low rumble, and a painting of George Washington, hanging on the wall, winks out of existence. Perhaps they have taken a souvenir.

Sally runs toward the hallway. "I'm getting out of here!"

"Sally, come back! I've made a deal with them. We're safe."

There, among some tall potted trees, Sally stops running. She stands perfectly still, perfectly straight. One hand grasps a tortured convolution of ficus bark, like talons of an eagle digging into the gray matter of a cat's cerebrum. Sally's hair blows in the slight breeze, and she beings to scream and scream and scream. . . .

"It's okay," you say and take her in your arms.

The president gazes all around like a madman. He suddenly realizes he is back in the White House and an expression of joy comes over him. Both Sally and the president look tired; their clothes are torn in places and covered with mud.

Narcinko salutes the president. "Glad to have you back, sir."

The president returns the salute.

You hear some shuffling sounds, and head toward the burlap bag. You give it a squeeze and feel several shapes, like a bag of oranges. You lift the bag up, remove the rope, find the exact location from which you disappeared, and with all your might toss it into the air.

When the bag finally comes down, you run for it and squeeze it. Nothing is inside.

The shuffling sounds disappear.

You look at the bag. "I think we caught a piece of one of their infants in the bag. Maybe it was in a region of space that had adhesive properties. Maybe the baby was lost and didn't know how get away. But now it's free."

Sally looks at you. "How did you know what to do?"

"From what I saw upsilon in space, I surmised that the Omegamorphs have a technological society and are intelligent. I noticed they had buildings, metal, and leather. They cultivated crops. Although *my* vision was limited to seeing cross sections of the Omegamorphs, *they* could see and understood the photos of you and the president that I brought along. They could also see my flares, just like a 2-D Flatlander waving a torch would visually attract our attention. I coaxed an Omegamorph down until its eye was on the same 'plane' with me. Perhaps that wasn't necessary, but I wanted to confirm that some creature was paying attention to me."

The president walks closer to you. "The fourth dimension is a nice place to visit, but I wouldn't want to live there. Our 2-D retina just doesn't see enough. But the 4-D beings gave me a message."

Sally stands beside you. "What message, sir?"

"They gave me specifications for building a 3-D retina that we can plug ourselves into. It will allow us to see things we can't even dream of. You two are to supervise this work immediately."

You take a breath. "But sir, even if we could build a working spherical retina, our minds would be unable to interpret the information. It would be like giving sight to a man blind from birth."

"You are right. But if we raise our children from birth with these retinas, their minds will adapt. With the aid of the Omegamorphs, the children will experience the fourth dimension."

Narcinko simply listens, folds his arms, inscrutable, except for the slight tremor of his eyebrows.

Sally shakes her head. "Sir, you're not suggesting we take out children's eyes and replace them with these?"

"Possibly, although we may be able to attach the new eyes to the optic nerves while leaving their original eyes in place. Possibly we could place 3-D retinas inside their current eyes, but that may limit the children. If they kept both old and new eyes, the children could flip switches to determine which eyes they want, but I doubt they would choose their more limited organs. They might want to remove their original eyes forever."

"That's sick," Sally says. "It would be morally indefensible. There are parallels in the cases of deaf children who can be fitted with artificial hearing organs, but such things are opposed by many in the deaf community."

"The beings will pull the children upsilon into their world and eventually teach the children how they can best use their new eyes. It would be unethical not to give children the new retinas." The president pauses. "If a child were born with eyes that saw no colors, wouldn't it be cruel to withhold treatment that allowed them to see color?"

Sally puts her hands on her hips. "That's not the same."

The president shakes his head. "We've no right to deny our species access to higher universes. This is the next step in humanity's evolution. Think of the new philosophies and ideas we would develop. This would be a boon to humankind. The procedure of implanting 3-D retinas will not cause the children pain. The children will still be human: they will still love their parents. They will play with friends. They'll just see more of reality."

Narcinko grins. "We could use them as all-seeing superspies."

Sally shakes her head and looks at the president. "I hope you're not thinking of this from a national security standpoint and using these kids as superspies."

The president takes a deep breath. "There may be the temptation, but I'll make sure laws are passed to respect the children's privacy, to make sure they respect ours, and to ensure that the kids are raised in a loving, supportive environment."

You snap your fingers. "Buy why would the Omegamorphs want to help us enter their world?"

"Apparently it came down a chain of command."

"Chain of command?"

The president nods. "A bargain they made. If they help us come up a dimension, then 5-D beings will provide the Omegamorphs with hyperspherical retinas that will allow them to glimpse the fifth dimension. Perhaps the Omegamorphs will eventually be brought into the fifth dimension. And so on and so on up the chain of dimensions."

"My God!" Sally says.

"Yes."

The president sits down on a Louis XIV chair next to the wall. "This is the way the dimensionally impoverished advance every several millennia. It will be magnificent."

You turn to Sally and sigh. "Well it looks like we can wrap up our investigations. We understand how the fourth dimension can generate strange phenomena. The president is safe. Humankind will grow one step closer to God."

"Wait," Sally says, "I just thought of something. These 4-D children will be able to spy on our every move. Nothing would be safe. Forget what I said about using them as spies. They could peek into bedrooms, peer into our guts, maybe even learn how to steal priceless valuables from safes if they learn more about the fourth dimension."

The president nods. "That's right. That's why you will not disband your FBI investigations. We'll have to be on the watch for mischief makers and pranksters as they learn to use their 3-D retinas. Ideally, the childrens' consciousness will be so elevated that they will have more important or interesting concerns than spying on you during a romantic interlude, or stealing money from safes. But we must be ever vigilant. Speaking of romantic interludes, Sally, do you have a date tonight?"

Sally tosses her head back and trails her hand through her hair. "Mr. President, I would never consider such a thing. You are my superior and I am a professional."

Narcinko smiles.

The president sighs and turns to you. "Doing anything tonight?"

"I know a good sushi place on Washington Ave."

"Let's do it."

You turn to Sally. "See you later. We'll continue some of our investigations tomorrow. Perhaps Narcinko could take you to Moe's Pub. What do you think Narcinko? Ever date an FBI agent?"

You smile at Sally.

She picks up a small marble bust of Abraham Lincoln and throws it at you.

Several years have passed since you helped the president return from the fourth dimension to the White House. An Omegamorph living in the L'Enfant Plaza Hotel, room 4D, grew attracted to Sally. However, Sally rebuffed the Omegamorph, explaining that she did not care for interspecies dating and was certainly not an easy pickup. The Omegamorph responded by giving Sally a 4-D rose and the following poem.

Reflections on a Tesseract Rose

Your mind and heart are of dimensions unbounded
and I can give you nothing save what you give to me,
your belief in the smallness of the world is unfounded,
but your imagination can set you free.
To my lovely Sally I give a tesseract rose,
grown in a garden you can't yet see,
Appearing and disappearing it kisses your nose

sometimes single or naught or three.
Let then your imagination wander
to the land where I grew this flower
and your heart yearn for new places,
and you will see me and love me in my many faces,
and know wonders your kind calls magical power,
and know riddles that even your Sphinx could not ponder.

Although touched by the being's show of emotion, Sally declined his romantic advances and eventually married Narcinko. A year later, you discovered that the man in the Santa Claus outfit was working for the Omegamorphs, helping people lift off into the fourth dimension and providing neural implants so explorers could see more clearly while taking transuniversal journeys.

A day ago you received your implants and decided to make a trip. Tanya, a woman who had already made several journeys into the fourth dimension, is by your side, functioning as a tour guide. After several months of intensive training with Tanya, studying ordinary shapes in your own world, you feel you are ready for the ultimate trip.

Occasionally Tanya disappears, wreathed in fog. It often seems that you are going in different directions. As you fly, you recall Sally's admonition not to worry about the phantom Santa Claus. At the time, it seemed there was no real need to find him. But perhaps Sally had, all along, been trying to tell you not to feel guilty about your dreams of flight. She knew it was important always to reach out, break the barriers, leave personal demons behind—to escape from your cages.

Tanya seems to lose altitude for a second, but it is probably only your imagination. For a dizzying minute you feel as if you are about to fall.

Oh no! Down you plunge. A windy sound fills your ears. Below, you think you see the Washington Monument, ominous, powerful, looming up like a bleached bone. But no. It is only your difficulty of adjusting to the new way of perceiving.

Space rushes by. Maybe stars and galaxies. And then there is the flashing of small creatures, many with tiny, symmetrical tiles on their surfaces (Fig. 6.4).

The sounds grow quieter, except . . . every now and then you think you hear a faint sigh, barely audible. At first you think it is Tanya, but you hear the sound repeatedly and finally realize it is coming from some-

Figure 6.4 Cross-sections of higher dimensional creatures.

where ahead of you, sometimes on one side, sometimes on the other. Suddenly there is a deep rushing whoosh close by and you see a dark body arcing through space.

"Was that an Omegamorph?" you say. The fleeting image lingers in your mind, like a vision of a mermaid.

"No," Tanya replies, "the body was bluish-gray. Omegamorphs have more fleshy tones."

No creature comes close again, though large fleeting images flash by, and the sighs continue along with faint squeaks and chirps.

"Something big ahead," Tanya says.

"Where?"

As you speak, you see something propel itself downward—out of sight as it makes a sighing sound.

"There," Tanya says and points her slender finger at two looming masses, huge, much bigger than elephants. They look like spaceships.

You slow your speed as two vessels swim toward you. But they aren't really vessels. Oh my God! They are huge manatees. You know all about manatees, but have never seen one up close.

On Earth, manatees are large water mammals, popularly called sea cows. Their dark bodies taper to flattened tails. Their forelimbs are flippers set close to the head; no external hind limbs exist. Their heads are small, with straight snouts and cleft upper lips with bristly hairs. These space manatees are similar to their Earthly counterparts.

"They're from the fifth dimension," Tanya says with a smile. "We are only seeing their intersection with this space."

As you gaze up at their bellies from below, the manatees resemble sentinels, guarding the path ahead of you. Then one manatee "swims" low so that you can see its eye looking down on you. Tanya grabs your hand tightly.

"They won't hurt us," she says to herself.

The manatee on the right seems particularly bold. Its body has wrinkled skin around the sides. The creature swims forward, hesitates for just a second, and then touches both you and Tanya with its flipper. The flipper seems to penetrate your chest slightly, but with no ill effect. It is like a ghost hand floating through a wall. Perhaps the manatee wants to make sure your body is real. There is nothing in the manatee's motions that is alarming. Indeed, there is something in the manatee's demeanor that inspires confidence. It has a graceful gentleness, a certain age-old ease.

The manatee near Tanya produces xylophone-like notes as it gently touches her long hair. These are the sounds audible to humans. You guess that the manatee is also emitting sounds inaudible to the human ear.

"They're singing!" Tanya says as her eyes dart back and forth from manatee to manatee.

The songs of the space manatees become more sonorous. You imagine the sounds traveling for hundreds of miles, resonating with the songs of great invisible beings swimming to distant dimensions in the darkness.

You wave to the manatees and turn to Tanya. "What do they want?"

The sounds suddenly die, replaced by two deep voices, speaking in unison. "Greetings, travelers. Welcome to our world." Balls of light and little rainbow arcs come from their eyes.

"Oh God!" Tanya cries. "They're so beautiful."

The manatees stare at you for several seconds, as if assessing you. Again, you can't help thinking the space manatees look ancient, full of wisdom, possessing mighty minds that you can never fully penetrate. The manatees also look kind. Their hyperspherical eyes are omniscient.

The manatees speak again. "Follow us so that we may lead you to higher dimensions in safety."

As the manatees slowly turn, a great tail goes swishing over your body.

"C'mon," Tanya says. "Let's follow them."

The manatees leave a stream of lights in their wake that you find easy to follow. Occasionally you see vortices spiraling away from you, like miniature tornadoes. You stay away from them, just in case they are sufficiently strong to affect your motion.

You point at other manatees emerging from great polyhedral domes that float in space. "Look at them!"

It seems that the new manatees are coming out of the domes to watch the beautiful lights emitted by the manatees in front of you. The brilliant colors form a dazzling display as the lights skim over the domes like shooting stars, rising high into space and then swooping gracefully to form chains of interlinked light trails.

Ahead is a 5-D city. The manatees never stop as they pass it by, but you turn your head so you can see the nearest building replete with manatee statues of translucent glass.

The building has a huge entry and is of colossal dimensions. Here and there other manatees are emerging from big open portals that yawn shadowy and mysterious.

You look up. The world rising above you is a tangle of strange architectures decorated with braids of 5-D flowers—or so you think. There are a number of tall spirals of stone that measure at least 100 yards in diameter at their bases. There are amber, pagoda-like plants, wonderfully tinted with red about their leaves. The plants seem to be part of the vast ancient architectural structures.

"Where are the manatees?" Tanya says.

"There." In the distance, you see the manatees separate by about thirty degrees as they disappear below some kind of 4-D horizon. A triplet of red orbs shoots across the sky and then skims along volcanic cliffs.

"Good-bye," the manatees say in their oboe-like voices, and they take off.

Although the space is filled with turbulent clouds left by the sudden disappearance of the manatees, there are clear breaks through which you can see new regions of space and vast crystalline cities that stretch as far as your eyes can see. Hyperspheres float like leaves in an autumnal breeze.

You grab Tanya's hand and point to the city. "Let's go!"

Tanya gives your hand a squeeze. You are no longer alone.

The crystal city is immersed in a soft tesseract of bright stars.

> A sphere, which is as many thousand spheres;
> Solid as crystal, yet through all its mass
> Flow, as through empty space, music and light;
> Ten thousand orbs involving and involved.
> Purple and azure, white green and golden,
> Sphere within sphere; and every space between
> Peopled with unimaginable shapes,
> Such as ghosts dream dwell in the lampless deep;
> Yet each inter-transpicuous; and they whirl
> Over each other with a thousand motions,
> Upon a thousand sightless axles spinning,
> And with the force of self-destroying swiftness,
> Intensely, slowly, solemnly, roll on,

Kindling with mingled sounds, and many tones,
Intelligible words and music wild.
With mighty whirl and multitudinous orb
Grinds the bright brook into an azure mist
Of elemental subtlety, like light.

   —Percy Bysshe Shelley, *Prometheus Unbound*

We are in the position of a little child entering a huge library whose walls are covered to the ceiling with books in many different tongues. . . . The child does not understand the languages in which they are written. He notes a definite plan in the arrangement of books, a mysterious order which he does not comprehend, but only dimly suspects.

—Albert Einstein

To understand the things that are at our door is the best preparation for understanding those things that lie beyond.

—Hypatia

And when He opened the seventh seal, there was silence in heaven about the space of half an hour.

—*Revelation* (8:1)

concluding remarks

THE HUBBLE TELESCOPE FINALLY SEES
THE BEGINNING OF THE UNIVERSE.

This completes our study of the fourth dimension. But in doing so, I wonder why I am personally so compelled to contemplate higher dimensions. It seems to me that it is our nature to dream, to search, and to wonder about our place in a seemingly lonely cosmos. Perhaps for this reason philosophers and even theologians have speculated about the existence of a fourth dimension and what its inhabitants might be like. I agree with Eric Fromm who wrote in *The Art of Loving*: "The deepest need of man is to overcome his separateness, to leave the prison of his aloneness."

The biggest question raised in this book is, "Can humans ever access a fourth dimension?" Or are we more like fish in a pond, near the surface, inches away from a new world, but forever confined, isolated by a seemingly impenetrable boundary: the pond's 2-D surface. If higher spatial dimensions exist, humanity may still have to wait a few hundred years before developing the capacity to explore them, but such a capacity may evolve for our survival—much like fish learned to leave the confines of their pools through evolution. It is possible that humans will some day have proof of higher spatial dimensions, such as those suggested by Kaluza-Klein theories. As discussed in Chapter 4, many cosmological models have been devised in which our universe curves through four-space in a way that could, in theory, be tested. For example, Einstein suggested a universe model in which a spaceship could set out in any direction and return to the starting point. In this model, our 3-D universe is treated as the hypersurface of a huge hypersphere. Going around it would be comparable to an ant's walking around the surface of a sphere. In other universe models, our universe is a hypersurface that twists through four-space like a Klein bottle. These are closed, one-sided, edgeless surfaces that twist on themselves like a Möbius strip. Using various satellites, astronomers now actively search for evidence of the universe's shape by studying temperature fluctuations in deep space.[1]

Can we learn to see the fourth dimension? Our inability to clearly visualize hyperspheres and hypercubes may result solely from our lack of training since birth. Our memories are of 3-D worlds, but who knows what might be accomplished with proper early training? This question has been discussed seriously by a number of mathematicians. Also, to say the fourth dimension is beyond imagination may be an exaggeration if we consider how far humans have stretched their imaginations since prehistoric time. From electrons to black holes, the history of science is the history of accepting concepts beyond our imagination. As Edward Kasner and James Newman note in *Mathematics and the Imagination*, "For primitive man to imagine the wheel, or a pane of glass, must have required even higher powers than for us to conceive of a fourth dimension." Whatever our limitations may be, even today the geometry of four dimensions is an indispensable part of mathematics and physics.

Worldview

The definitive discovery of 4-D beings would drastically alter our worldview and change our society as profoundly as the Copernican, Darwinian, and Einsteinian revolutions. It would impact religions and spur interest in science as never before.

If intelligent 4-D beings evolved and we were able to communicate with them, our correspondence could bring us a richer treasure of information than medieval Europe inherited from ancient Greeks like Plato and Aristotle. Just imagine the rewards of learning a 4-D being's language, music, art, mythology, philosophy, biology, even politics. Who would be their mythical heroes? Are their Gods more like the thundering Zeus and Yaweh or the gentler Jesus and Baha'u'llah?

Wouldn't it be a wild world in which to live if 4-D devices were as common as the computer and telephone? In such a world, it might be possible to manipulate space and time to make travel to other worlds easier. Mathematicians dating back to Georg Bernhard Riemann have studied the properties of multiply connected spaces in which different regions of space and time are spliced together. Physicists, who once considered this an intellectual exercise for armchair speculation, are now seriously studying advanced branches of mathematics to create practical models of our universe and better understand the possibilities of parallel worlds and travel using wormholes and manipulating time.

Zen Buddhists have developed questions and statements called *koans* that function as a meditative discipline. Koans ready the mind so that it can entertain new intuitions, perceptions, and ideas. Koans cannot be answered in ordinary ways because they are paradoxical; they function as tools for enlightenment because they jar the mind. Similarly, the contemplation of 4-D life is replete with koans; that is why these final paragraphs pose more questions then they answer. These questions are koans for scientific minds.

Hyperspace Survival

As our technology advances, perhaps one day the fourth dimension—and hyperspace connections to other regions of space—will provide a refuge for humans as their Sun dies. The Earth is like an inmate waiting on death row. Even if we do not die from a comet or asteroid impact, we know the Earth's days are numbered. The Earth's rotation is slowing down. Far in the future, day lengths will be equivalent to fifty of our present days. The Moon will hang in the same place in the sky and the lunar tides will stop.

In five billion years, the fuel in our Sun will be exhausted, and the Sun will begin to die and expand, becoming a red giant. At some point, our oceans will boil away. No one on Earth will be alive to see a red glow filling most of the sky. As Freeman Dyson once said, "No matter how deep we burrow into the Earth . . . we can only postpone by a few million years our miserable end."

Where will humans be, five billion years from now, at the end of the world? Even if we could somehow withstand the incredible heat of the sun, we would not survive. In about seven billion years, the Sun's outer "atmosphere" will engulf the Earth. Due to atmospheric friction, the Earth will spiral into the sun and incinerate. In one trillion years, stars will cease to form and all large stars will have become neutron stars or black holes. In 100 trillion years, even the longest-lived stars will have used up all their fuel.

If this ending seems too dismal, perhaps we should ask if there is hope for humanity when the Sun expands to engulf the Earth in seven billion years. To give an answer, first consider that around four billion years ago, living creatures were nothing more than biochemical machines capable of self-reproduction. In a mere fraction of this time, humans evolved from creatures like Australopithecines. Today humans have wandered the Moon and have studied ideas ranging from general relativity to quantum cosmology. Once space travel begins in earnest, our descendents will leave the confinement of Earth. Because the ultimate fate of the universe involves great cold or great heat, it is likely that *Homo sapiens* will become extinct. However, our civilization and our values may not be doomed. Who knows into what beings we will evolve? Who knows what intelligent machines we will create that will be our ultimate heirs? These creatures might survive virtually forever. They may be able to easily contemplate higher dimensions, and our ideas, hopes, and dreams carried with them. There is a strangeness to the loom of our universe that may encompass time travel, higher dimensions, quantum superspace, and parallel universes— worlds that resemble our own and perhaps even occupy the same space as our own in some ghostly manner.

Some physicists have suggested that the fourth dimension may provide the only refuge for intelligent life. Michio Kaku, author of *Hyperspace*, suggests that "in the last seconds of death of our universe, intelligent life may escape the collapse by fleeing into hyperspace." Our heirs, whatever or whoever they may be, will explore these new possibilities. They will explore space and time. They will seek their salvation in the higher universes.

The upbeat feelings of theoretical physicist Freeman J. Dyson best express my beliefs:

> Gödel proved that the world of pure mathematics is inexhaustible; no finite set of axioms and rules of inference can ever encompass the whole of mathematics; given any finite set of axioms, we can find meaningful mathematical questions which the axioms leave unan-

swered. I hope that an analogous situation exists in the physical world. If my view of the future is correct, it means that the world of physics and astronomy is also inexhaustible; no matter how far we go into the future, there will always be new things happening, new information coming in, new worlds to explore, a constantly expanding domain of life, consciousness, and memory.

Let's end this book at a point from where we started. Throughout, I have referred to clergyman Edwin Abbott Abbott's 2-D world called Flatland. The book describing it is both scientific and mystical. I wonder if Abbott actually thought that the fourth dimension was the key to understanding God. I am doubtful. In his book *The Spirit on the Waters*, written nearly ten years after *Flatland*, Abbott recounts *Flatland's* climactic scene in which the 2-D hero is confronted by the changing shapes of a 3-D being as it passes through Flatland. The Flatlander does not worship this being because of its God-like powers. Rather, Abbott suggests that miraculous powers do not necessarily signify any of the moral and spiritual qualities required for worship and adoration. Abbott concludes:

> This illustration from four dimensions, suggesting other illustrations derivable from mathematics, may serve a double purpose in our present investigation. On the one hand it may lead us to vaster views of possible circumstances and existence; on the other hand it may teach us that the conception of such possibilities cannot, by any direct path, bring us closer to God. Mathematics may help us to measure and weigh the planets, to discover the materials of which they are composed, to extract light and warmth from the motion of water and to dominate the material universe; but even if by these means we could mount up to Mars or hold converse with the inhabitants of Jupiter or Saturn, we should be no nearer to the divine throne, except so far as these new experiences might develop in our modesty, respect for facts, a deeper reverence for order and harmony, and a mind more open to new observations and to fresh inferences from old truths.

Abbott believed that study of the fourth dimension is important in expanding our imagination, increasing our reverence for the Universe, and increasing our humility—perhaps the first steps in any attempt to understand the mind of God.

Mulder: I've seen too many things not to believe.

Scully: I've seen things too. But there are answers to be found now. We have hope that there is a place to start. That's what I believe.

Mulder: You put such faith in your science, Scully, but . . . for the things I have seen science provides no place to start.

Scully: Nothing happens in contradiction to nature, only in contradiction to what we know of it, and that's a place to start. That's where the hope is.

—"Herrenvolk," *The X-Files*

appendix a

mind-bending four-dimensional puzzles

Our cosmos—the world we see, hear, feel—is the three-dimensional "surface" of a vast, four-dimensional sea. . . . What lies outside the sea's surface? The wholly other world of God! No longer is theology embarrassed by the contradiction between God's imminence and transcendence. Hyperspace touches every point of three-space. God is closer to us than our breathing. He can see every portion of our world, touch every particle without moving a finger though our space. Yet the Kingdom of God is completely "outside" of three-space, in a direction in which we cannot even point.

—Martin Gardner, "The Church of the Fourth Dimension"

Death is a primitive concept; I prefer to think of them as battling evil—in another dimension!

—Grig in *The Last Starfighter*

I couldn't help including a few mind-stretching puzzles in this book, although most would be considered too difficult to solve even by Ph.Ds in mathematics. Nevertheless, reading through the questions and solutions should be sufficiently mind-expanding to justify their inclusion. Similar kinds of puzzles have been discussed in my earlier books, such as *The Alien IQ Test* and *Mazes for the Mind*.

Lost in Hyperspace

Aliens abduct you and place, in front of your paralyzed eyes, a twisting tube within which a robotic ant crawls. The ant starts at a point marked by a glowing red dot. The aliens turn to you and say, "This ant is executing an infinite random walk; that is, it walks forever by moving randomly one step forward or one step back in the tube. Assume that the tube is infinitely long. What is the probability that the random walk will eventually take the ant back to its starting point?" You have one week to answer correctly or else the aliens will examine your internal organ systems with a pneumoprobe.

You have all the information you need to solve this problem. The ant essentially lives in a 1-D universe. How would your answer change for higher dimensions?

For a solution, see note 1.

A Tesseract in the FBI Headquarters

Aliens have descended to Earth and placed a 3 × 3 × 3 × 3–foot Rubik's tesseract at the FBI headquarters in Washington, D.C. (A tesseract is a 4-D cube in the same way that a cube is a 3-D version of a square.) The colors of this 4-D Rubik's cube shift every second for several minutes as onlookers stare and scream. (At first the FBI believes it is a Russian spy device.) Finally, the tesseract is still—permitting us to scramble it by twisting any of its eight cubical "faces," as described below. You are sent to investigate.

You soon realize that this is an alien test, and humans have a year to unscramble the figure, or Washington, D.C., will be annihilated. Your question: What is the total number of positions of the tesseract? Is the number greater or less than a billion? For a solution, see note 2.

Background to a Four-Dimensional Rubik's Cube

Many of you will be familiar with Ernö Rubik's ingenious, colorful 3 × 3 × 3 cubical puzzle (Fig. A.1). Each face is a 3 × 3 arrangement of small cubes called "cubies." If you were to cut this cube into three layers, each layer would look like a 3 × 3 square with the same four colors appearing along its sides. Two additional colors are in the interiors of all the squares in the first and third layers. (These are the colors on the bottom of all the squares in the first layer and top of all the squares in the third layer.)

The aliens have extended this puzzle to the fourth dimension where the 4-D 3 × 3 × 3 × 3 Rubik's hypercube, or tesseract, is composed of 3 × 3 × 3 cubes stacked up in the fourth dimension. All cubes have the same six colors assigned to their faces; in addition, two more colors are assigned to the interiors of all the little cubes (cubies) in the first and third cube. (I refer to the eighty-one small cubes in this representation as cubies, as have other researchers like Dan Velleman, although each is really one of the eighty-one small tesseracts that make up the alien Rubik's tesseract.)

The 3-D and 4-D puzzles differ in the following ways. The original Rubik's cube has six square faces. The Rubik's tesseract has eight cubical "faces." In the standard Rubik's cube, there are three kinds of cubies: edge cubies with two colors, corner cubies with three colors, and face-center cubies with one color. (I ignore the

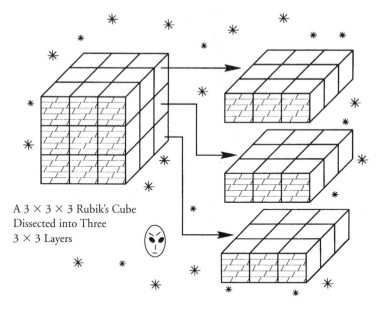

A 3 × 3 × 3 Rubik's Cube
Dissected into Three
3 × 3 Layers

Figure A.1   A 3 × 3 × 3 Rubik's cube dissected into three 3 × 3 layers.

cubie in the center of the cube that has no color and plays no role in the puzzle.) Rubik's tesseract has four kinds of pieces that are also distinguished by the number of their colors.

Those of you with computers may enjoy "MagicCube4D," a fully functional 4-D analog of Rubik's cube developed by Daniel Green and Don Hatch. The current implementation is for Windows95 and Windows NT. The graphical model is an exact 4-D extension of the original, plastic 3-D puzzle, but with some useful features such as a "reset" button. Employing the same mathematical techniques that are used to project 3-D objects onto 2-D screens, MagicCube software "projects" the 4-D cube into three dimensions. The resulting 3-D objects can then be rendered with conventional graphics software onto the screen. It is very difficult to solve the 4-D Rubik's cube starting from a scrambled initial structure. If you ever do succeed, you will be one of a *very* elite group of people. You will almost certainly need to have previously mastered the original Rubik's cube before you can hope to solve this one. Luckily, all the skills learned for the original puzzle will help you with this one. Also, you don't need to ever solve the full puzzle to enjoy it. One fun game is to start with a slightly scrambled configuration, just a step or two away from the solved state, and work to back out those few random twists. If you get tired trying to solve the puzzle yourself, it is breathtaking to

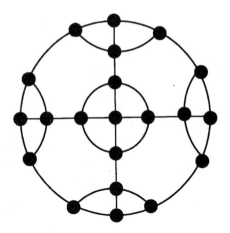

Figure A.2 Initial empty Sz'kwa board.

watch the computer do it. For more information, visit: http://www.superliminal.com/cube.htm.

Hyperdimensional Sz'kwa

An alien challenges you to a simple-looking, competitive game. You hold twenty-five white crystals in your hand; the alien holds twenty-five black crystals. At the start of the game, the circular board would have no crystals (Fig. A.2).

You and the alien take turns placing a crystal on the board at the positions with black dots. The rules are as follows. If a player's crystal is completely surrounded by the opponent's crystals, it is captured. Figure A.3 shows the capture of a black piece (left) and the capture of two white pieces (right). When a player has no crystals left to place on the circular board, or no empty sites on which to place a crystal without its being captured, the game ends. The winner is the player who holds the greatest number of crystals.

How many different arrangements of crystals on the playing board exist? Is it better to be the first player? Can you write a program that learns strategies by playing hundreds of games and observing its mistakes? Develop a multidimensional Sz'kwa game where the center site on the Sz'kwa board connects center sites on adjacent boards or where all sites connect to the corresponding positions on adjacent boards. First try a game using just two connected boards and then three. Generalize your discoveries to *N* boards. These versions represent hyperdimensional Sz'kwa (Fig. A.4).

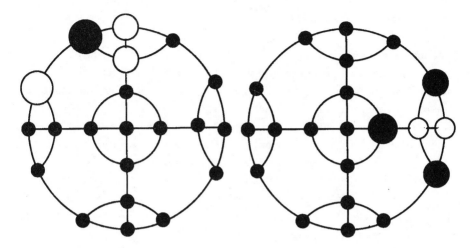

Figure A.3   The capture of a black piece (left), and the capture of two white pieces (right).

Figure A.4   Hyperdimensional Sz'kwa.

Hypertetrahedral Numbers

Tetrahedral numbers form the sequence: 1, 4, 10, 20, 35, 56, 84, 120 . . . with a generating formula $(1/6)n(n + 1)(n + 2)$. This can be best visualized using cannonballs in a pyramid-shaped pile with a triangular base. Starting from the top of the pile, the number of balls in each layer is 1, 3, 6, 10, 15, . . . , which forms a sequence of triangular numbers because each level is shaped like a triangle. Tetrahedral numbers can be thought of as sums of the triangular numbers. We can extend this idea into higher dimensions and into hyperspace. In 4-D space, the piles of tetrahedral numbers can themselves be piled up into 4-D, hypertetrahedral numbers: 1, 5, 15, 35, 70. . . . We can form these numbers from the general formula: $(1/24)n(n + 1)(n + 2)(n + 3)$. Can you impress your friends by generating hyper-hypertetrahedral numbers?

appendix b

higher dimensions in science fiction

Whether we will ultimately be able to create furniture from curved space, partake of a multidimensional reality, or directly view all of humanity's alternate histories, becomes less of a issue than being able to fuel the imagination with these endless possibilities.

—Sten Odenwald, from the Internet

The children were vanishing. They went in fragments, like thick smoke in a wind, or like movement in a distorting mirror. Hand in hand, they went, in a direction Paradine could not understand. . . .

—Lewis Padgett, "Mimsy Were the Borogoves"

In this appendix, I list fascinating science-fiction stories and novels that deal with the fourth dimension. Where possible, I have attempted to provide the publisher for each book. (Unfortunately, in some cases, I was not able to find publishers for out-of-print books.) A number of the science-fiction books were suggested by Dr. Sten Odenwald, author of *The Astronomy Cafe*. Dr. Odenwald received his Ph.D. in astronomy from Harvard University and maintains several interesting web pages such as "Ask the Astronomer" at http://www2.ari.net/home/odenwald/qadir/ qanda.html. Note that *Fantasia Mathematica*, cited in many references, has recently been republished by Springer-Verlag under the Copernicus imprint.

Many books in the science-fiction genre discuss space travel through hyperspace, but they are too numerous to list. Many deal with ways in which spaceships take shortcuts to get quickly from one point in the universe to another. Hyperspace is often described as a higher-dimensional space through which our 3-D space can be folded so that two seemingly distant points are brought close together. Famous books dealing with hyperspace include: Isaac Asimov's *Foundation* series, Larry Niven's *The Borderland of Sol* (1974), Jerry Pournelle's *He Fell into a Dark Hole* (1974), Larry Niven's and Jerry Pournelle's *A Mote in God's Eye*, Arthur J. Burk's *The First Shall be Last*, Raymond F. Jones' *Correspondence Course* (1945), and Carl Sagan's *Contact* (1997) which describes a machine for creating a "dimple in spacetime" and to which alien engineers attach a "wormhole" bridge.

There are also many books that discuss alternate or parallel universes. Often a parallel world is another universe situated alongside our own, displaced along a fourth spatial dimension. Parallel worlds are often referred to in science fiction as "other dimensions." Modern use of the term "parallel universes" often employs an infinite number of parallel worlds containing all possible Earthly histories. The notion that the universe is one single aspect of such a "multiverse" has gained some acceptance by the "many-worlds interpretation" of quantum mechanics. As an example, Robert Heinlein's science-fiction novel *The Number of the Beast* discusses a parallel world that appears identical to ours in every respect except that the letter "J" does not appear in the English language. Similarly, in A. E. van Vogt's *Recruiting Station* (1942), future humans recruit twentieth-century people to fight a war. The terrestrial time stream is manipulated to create eighteen alternate solar systems in which battles over political control are waged.

1. Abernathy, R. (1950) "The Ultimate Peril." Venusian psycho-physicists attacking Earth with hyperspace weapons.

2. Archette, G. (1948) "Secret of the Yellow Crystal." Ancient Martians rearrange the molecular structure of crystals, without mechanical technology, to tap "extra-dimensional or sub-spatial energies." They use hyperspace to leave Mars rather than face extinction. (Guy Archette is a pseudonym for Chester Geier.)

3. Asimov, I. (1947) "Little Lost Robot." Describes the use of "Hyperatomic Drive" shortened to "Hyperdrive." "Fooling around with hyper-space isn't fun. We run the risk of blowing a hole in normal space-time fabric and dropping right out of the universe."

4. Asimov, I. (1992) *The Stars Like Dust.* New York: Spectra. (Originally published in 1950.) Describes travel in hyperspace.

5. Asimov, I. (1990) *The End of Eternity.* New York: Bantam Books. Future humans live outside normal spacetime and fix the past in order to facilitate human progress. This story spans millions of centuries. Computers aid technicians in making subtle adjustments. A barrier in spacetime is discovered millions of years in the future—caused by even more advanced humans protecting themselves from the world-line tampering going on during these earlier ages.

6. Bear, G. (1995) *Eon.* New York: Tor. Humans discover a hundred-kilometer-long asteroid that extends inside billions of miles! Built by humans thirteen centuries in our future, the asteroid is fashioned out of artificially twisted spacetime and functions as a tunnel into hyperspace. As one travels down

the asteroid's axis, time is shifted. Alternate universes are stacked within each millimeter of the tunnel.

7. Bester, A. (1942) "The Push of a Finger." *The Astounding-Analog Reader, Volume One*, Harry Harrison and Brian W. Aldiss, eds. New York: Doubleday. Scientists create an "osmotic spatial membrane" to tap energy from hyperspace. Unfortunately, this energy begins to drain into our universe, causing it to come to a premature end. Fortunately, this is stopped by a time traveler from the future who intervenes in the nick of time.

8. Bond, N. (1974) "The Monster from Nowhere," in *As Tomorrow Becomes Today*, Charles W. Sullivan, ed. New York: Prentice-Hall. (Originally published in *Fantastic Adventures*, July 1939.) Humans trap a 4-D creature in our world. Also see Nelson Bond's 1943 short story "That Worlds May Live" that describes hyperspace propulsion systems. Bond describes a "qaudridimensional drive," the first artificial space warp into the fourth dimension—created by Jovian scientists. "The Jovians create a four-dimensional space warp between points in three-dimensional space. A magnetized flux field warps three-dimensional space in the direction of travel. . . . It's as easy as that."

9. Bond, N. (1950) "The Scientific Pioneer Returns," in *Lancelot Biggs: Spaceman* by Nelson S. Bond. New York: Doubleday. (Story originally published in 1940.) A ship accelerates into "imaginary space" that turns out to be a parallel universe. "Einstein and Planck fiddled around with hyperspatial mechanics and discovered that mass is altered when it travels at high velocity. The gadget worked better than you expected."

10. Brueuer, M. (1958) "The Captured Cross-Section," in *Fantasia Mathematica*, C. Fadiman, ed. New York: Simon and Schuster. A young mathematician chases his fiancée into the fourth dimension. An excellent physical description of 4-D creatures caught in our own world.

11. Brunner, J. (1985) *Age of Miracles*. New York: New American Library. Earth is invaded by a dozen "cities of light." Their interiors are twisted into higher dimensions and result in disturbing sensory shifts to any unshielded human who enters. Humans are able to use these portals without being noticed by the aliens.

12. Campbell, J. (1934) "The Mightiest Machine." John Campbell coins the word "hyperspace" in this story.

13. Clarke, A. C. (1956) "Technical Error." A technician is rotated through the fourth dimension and becomes reversed—a condition in which he can no longer metabolize food unless it is provided to him in the "left-handed" state.

14. Cramer, J. (1989) *Twistor*. New York: William Morrow. Cramer is a professor of physics at the University of Washington. The protagonists in the book, a male postdoc and a female graduate student working in the University of Washington physics department, are conducting experiments that use a peculiar configuration of electromagnetic fields that rotates normal matter into shadow matter (predicted by string theory) and vice versa, rotating a stage where one set is replaced by another. The first time this happens, a spherical volume containing expensive equipment disappears. Subsequently, the postdoc and two small children are "rotated" to a shadow-matter Earth and trapped inside a huge tree; an enormous sphere replaces them in the middle of their Seattle laboratory.

15. Deutsch, A. (1958) "A Subway Named Moebius," in *Fantasia Mathematica*, C. Fadiman, ed. New York: Simon and Schuster. A Harvard professor of mathematics is asked to solve a mysterious disaster in Boston's underground transportation system.

16. Egan, G, (1995) *Quarantine*. New York: HarperCollins. The human mind creates the universe it perceives by quantum-mechanically destroying all other possible universes. The book's characters (and the readers) are forced to ask what is real.

17. Gamow, G. (1962) "The Heart on the Other Side," in *The Expert Dreamers*, F. Pohl, ed. New York: Doubleday.

18. Gardner, M. (1958) "No-Sided Professor," in *Fantasia Mathematica*, C. Fadiman, ed. New York: Simon and Schuster. Describes what happens when a professor of topology meets Dolores, a striptease artist.

19. Geier, C. (1954) "Environment," in *Strange Adventures in Science Fiction*, Groff Conklin, ed. New York: Grayson. (Originally published in 1944.) Uses the term "hyperspacial drive." "You go in here, and you come out there . . . ."

20. Geier, S. (1948) "The Flight of the Starling." A spaceship circumnavigates the solar system in three hours using atomic-powered warp generators. These generators "create a warp in space around the ship . . . a moving ripple in the fabric of space." The ship rides this ripple like a surfer on an ocean wave. Between normal space and negative space is a zone called hyperspace. In negative space, time travel is possible.

21. Hamling, W. (1947) "Orphan of Atlans." A natural cataclysm unleashes forces and "a rent is made in the ether itself. . . . A great space warp forms around Atlantis." This catapults the last few survivors of Atlantis out of their normal spacetime and into the twentieth century.

22. Heinlien, R. (1991) *Starman Jones*. New York: Ballantine. Spaceships travel through 4-D hyperspace.

23. Heinlein, R. (1987) *Citizen of the Galaxy*. New York: Ballantine. (Originally published in 1957.) Travel in hyperspace.

24. Heinlein, R. (1958) "—And he built a crooked house," in *Fantasia Mathematica*, C. Fadiman, ed. New York: Simon and Schuster. (Original story published in 1940.) The misadventures of a California architect who built his house to resemble the projection in 3-D space of a 4-D hypercube. When the hypercube house folds, it looks like an ordinary cube from the outside because it rests in our space on its cubical face—just as a folded paper cube, sitting on a plane, would look to Flatlanders like a square. Eventually the hypercube house falls out of space altogether.

25. Heinlein, R. (1991) *Starman Jones*. New York: Ballantine. (Originally published in 1953.) The transition into "*N*-space" requires careful calculations because at some points in interstellar space, space is folded over on itself in "Horst Anomalies."

26. Laumer, K. (1986) *Worlds of the Imperium*. New York: TOR. Describes alternate worlds in which one man is confronted by his alternate self. The protagonist is trapped and kidnapped by the inhabitants of a parallel universe. He learns that he must assassinate a version of himself who is an evil dictator in the parallel world.

27. Leiber, F. (1945) *Destiny Times Three*. New York: Galaxy Novels. Several humans use a "Probability Engine" to split time and create alternate histories, allowing only those best suited for Earth to survive.

28. Lenz, F. (1997) *Snowboarding to Nirvana*. New York: St. Martin's Press. Frederick Lenz is introduced to "skyboarding" in a higher dimension. He proceeds to skyboard through colored dimensions until he reaches a violet one. "The air in this dimension was textured with some kind of indecipherable hieroglyphic writing. Beings like huge American Indians began flying past us."

29. Lesser, M. (1950) *All Heroes Are Hated*. In the year 2900 A.D., interstellar travel is commonplace and quick using hyperspace. Unfortunately, twelve billion inhabitants of six worlds are annihilated when a spaceship exits hyperspace with its drive still turned on.

30. Martin, G. R. R. (1978) "FTA," in *100 Great Science Fiction Short Stories*. I. Asimov, M. Greenberg, and J. Olander, eds. New York: Doubleday. Hyperspace turns out not to be a shortcut for space travel.

31. Moorcock, M. (1974) *The Blood Red Game.* New York: Mayflower Books. (Originally published in *Science Fiction Adventures* as *The Sundered Worlds*, 1962.) Humans navigate through hyperspace and the other "alien dimensions." The hero of the story, Renark, encounters odd humanoid beings that tell him that the universe will recollapse in a year. To save humanity, Renark must find a way for humans to leave our 4-D spacetime (three dimensions of space, one dimension of time). He travels to a solar system whose orbit is at right angles to the rest of spacetime and passes through our universe every few hundred years. Our 4-D (spacetime) universe coexists with an infinite number of other universes. In this multiverse theory, our universe is like a page in a book; each page has its own physical laws and beings. In this strange solar system, Renark meets the Originators, multidimensional beings that developed and maintained the multiverse in order to create a nursery for a life-form to replace them and keep reality from falling apart.

32. Nourse, A. (1963) "Tiger by the Tail," in *Fifty Short Science Fiction Tales*, Isaac Asimov and Groff Conklin, eds. New York: Macmillan (Story originally published in 1951.) Creatures in the fourth dimension coerce a human shoplifter to send them aluminum through an interdimensional gateway resembling a pocketbook. The shoplifter is apprehended by police who realize the pocketbook's purpose. Lowering a hook into the pocketbook, the police manage to "pull a non-free section of their universe through the purse, putting a terrific strain on their whole geometric pattern. Their whole universe will be twisted." Now humanity has a ransom against invasion.

33. Padgett, L. (1981) "Mimsy Were the Borogroves," in *The Great SF Stories 5*, I. Asimov and M. Greenberg. New York: DAW. (Story originally published in 1943.) Paradine, a professor of philosophy, cannot understand where children are disappearing to. Earlier, the children find a wire model of a tesseract (4-D cube), with colored beads that slide along the wires in strange ways. It turns out that the model is a toy abacus that had been dropped into our world by a four-space scientist who is building a time machine. This abacus teaches the children how to think four dimensionally, and they finally walk into the fourth dimension and disappear.

34. Phillips. R. (1948) "The Cube Root of Conquest." We coexist along with other universes in space, but are separated in time.

35. Pohl, F. (1955) "The Mapmakers." Describes hyperspace as a pocket universe.

36. Pohl, F. (1987) *The Coming of Quantum Cats*. New York: Bantam. A parallel worlds story.

37. Pohl, F. and Williamson, J. (1987) *The Singers of Times*. New York: Ballantine. A parallel worlds story.

38. Schachner, N. (1938) "Simultaneous Worlds." All 3-D matter extends into a higher dimension. A machine images these higher worlds that resemble Earth.

39. Shaw, R. (1967) *Night Walk*. New York: Banner Books. A hyperspace universe has a fiendishly complicated shape.

40. Shaw, B. (1986) *The Two-Timers*. A man who lost his wife inadvertently creates a parallel world in which she still exists.

41. Shaw, B. (1987) *A Wreath of Stars*. New York: Baen Books. Two worlds made of different kinds of matter coexist until the approach of a star shifts the orbit of one of them.

42. Silverberg, R. (1972) *Trips*. Transuniversal tourists wander aimlessly through worlds with varying similarities.

43. Simak, C. (1992) *Ring Around the Sun*. New York: Carroll & Graf. (Originally published in 1953.) Describes a series of Earths, empty of humanity and available for colonization and exploitation.

44. Simak, C. (1943) *Shadow of Life*. Martians learn to shrink themselves to subatomic size by extending into the fourth dimension, causing them to lose mass and size in the other three dimensions.

45. Smith, E. E. (1939) *Grey Lensman*. A crew feels as though they "were being compressed, not as a whole, but atom by atom . . . twisted . . . extruded . . . in an unknowable and non-existent direction. . . . Hyperspace is funny that way. . . ." A weapon known as a "hyperspatial tube" is used. It is described as an "extradimensional" vortex. The terminus of the tube cannot be established too close to a star due to the tube's apparent sensitivity to gravitational fields.

46. Smith, M. (1949) *The Mystery of Element 117*. Describes how our universe extends a short distance into a fourth spatial dimension. Because of this, it is possible to rotate matter completely out of three-space by building a 4-D translator. Element 117 opens a portal into this new dimension inhabited by humans who have died. They live in a neighboring world to ours, slightly shifted from ours along the fourth dimension. "Our 3-space is but one hyperplane of hyperspace." Succeeding layers are linked together like pages in a book.

47. Tenneshaw, S. (1950) "Who's That Knocking at My Door?" A honeymooning couple's hyperdrive breaks down near a white dwarf star.

48. Upson, W. (1958) "A. Botts and the Moebius Strip," in *Fantasia Mathematica*, C. Fadiman, ed. New York: Simon and Schuster. A simple demonstration in topology saves the lives of several Australian soldiers.

49. van Vogt, A. E. (1971) "M 33 in Andromeda," in *M 33 in Andromeda*, A. E. van Vogt. New York: Paperback Library. (Originally published in 1943.) Humans receive mental messages from an advanced civilization in the Andromeda galaxy. Earthlings use "hyperspace" in planet-to-planet matter transmission. Focusing a hyperspace transmitter on a spaceship moving faster than light requires specifying coordinates in a 900,000-dimensional space.

50. Wandrei, D. (1954) "The Blinding Shadows," in *Beachheads in Space*, August Derleth, ed. New York: Weidenfeld Nicolson. (Story originally published in 1934.) An inventor named Dowdson builds a machine that sucks energy into the fourth dimension. In a speech to his colleagues, Dowdson remarks:

> Gentlemen, there was a time long ago when objects were considered to have two dimensions, namely, length and breadth. After Euclid, it was discovered that length, breadth, and thickness comprised three dimensions. For thousands of years, man could visualize only two dimensions, at right angles to each other. He was wrong. Now, for more thousands of years, man has been able to visualize only three dimensions, at right angles to each other. May there not be a fourth dimension, perhaps at right angles to these, in some fashion that we cannot yet picture, or perhaps lying altogether beyond our range of vision? Objects emitting infra-red rays, and lying in such a four-dimensional world, might easily be past our ability to see and our capacity to understand, while existing beside us, nay, in this very hall.

The audience pays deep attention to Dowdson as he reaches the main point of his paper:

> A three-dimensional object casts a two-dimensional shadow. If such a thing as a two-dimensional object existed, doubtless it would throw a one-dimensional shadow. And should a four-dimensional solid be extant, its shadow would be three-dimensional. In other words, gentlemen, it is entirely conceivable

that in our very midst lies a four-dimensional world whose shadow would be characterized by three dimensions, though we might never have eyes to see or minds to understand the nature of the four-dimensional origin of that shadow.

Later in his paper, Dowdson states:

You may well ask why, if my theories are correct, no such shadow has ever been seen. The answer, I think, is fairly simple. Subject to laws alien to those we know, and imperceptible to our range of vision, it is quite probable that the object does cast such a shadow, but of such a color as to be also invisible. The alternative theory is that some intermediary, such as a mirror based upon radical principles, would reflect the shadow.

Dowdson invents a machine whose lenses not only rotate in three dimensions but also in the fourth dimension. He wants to capture images of objects in the fourth dimension in 3-D space. Unfortunately, the black shadows of alien beings soon appear and begin to consume the inhabitants of New York City:

The area involved roughly comprises what was formerly known as Greater New York, and includes a circle whose radius is some ten miles, even extending out into the harbor and the Atlantic. This area, now protected on land by great cement, steel, and barbed-wire fortifications erected by the government, is dead ground, which tens of thousands of sight-seers visit weekly to view the "lost" city and its strange conquerors, the Blinding Shadows.

Why they remain and what they seek are unsolved riddles, nor indeed is there surety that somewhere, sometime, they may not flame outside the barriers and sweep onward, or that some other scientist may not unwittingly loose upon the rest of the world a horde of mysterious, ravenous, and Blinding Shadows, against which mankind is powerless and about whose source nothing is known.

For ten years, the Blinding Shadows have possessed the dream city; and ten thousand times that many years are likely to slip into oblivion without one human tread in streets where not even the ravens hover and where the hellish Shadows endlessly rove.

51. Wells, H. G. (1952) "The Plattner Story," in *28 Science Fiction Stories.* New York. Dover. A mysterious green powder blows a young chemistry teacher named Plattner into the fourth dimension. Naturally, the students in the classroom are shocked to find that, when the smoke clears from the explosion, Plattner is gone. There is no sign of him anywhere. The school's principal has no explanation. During the nine days in four-space, Plattner sees a large green sun and unearthly inhabitants. When Plattner returns to our world, his body is reversed. His heart is now on the right and he writes mirror script with his left hand. It turns out that the quiet, drifting creatures in Wells's; four-space are the souls of those who once lived on Earth.

banchoff klein bottle

It is true that we are all at every moment situated simultaneously in all the spaces which together constitute the universe of spaces; for whenever there is disclosed to us the existence of a space which had previously been concealed from us, we know from the very first moment that this space has not just come into being, but that it had always surrounded us without our noticing it. Yet, nevertheless, we are not ourselves able to force open the gate which leads to a space that has so far been closed to us.

—Karl Heim, *Christian Faith and Natural Science*

In the last decade, even serious mathematicians have begun to enjoy and present bizarre mathematical patterns in new ways—ways sometimes dictated as much by a sense of aesthetics as by the needs of logic. Moreover, computer graphics allow nonmathematicians to better appreciate the complicated and interesting graphical behavior of simple formulas.

This appendix provides a recipe for creating a beautiful graphics gallery of mathematical surfaces. To produce these curves, I place spheres at locations determined by formulas that are implemented as computer algorithms. Many of you may find difficulty in drawing shaded spheres; however, quite attractive and informative figures can be drawn simply by placing colored dots at these same locations. Alternatively, just put black dots on a white background. As you implement the following descriptions, change the formulas slightly to see the graphic and artistic results. Don't let the complicated-looking formulas scare you. They're very easy to implement in the computer language of your choice by following the computer recipes and computational hints given in the program outlines.

Unlike the curves you may have seen in geometry books (such as bullet-shaped paraboloids and saddle surfaces) that are simple functions of x and y, certain surfaces occupying three dimensions can be expressed by *parametric equations* of the form: $x = f(u,v)$, $y = g(u,v)$, $z = h(u,v)$. This means that the position of a point in the third dimension is determined by three separate formulas. Because f, g, and h can be anything you like, the remarkable panoply of art forms made possible by plotting these surfaces is quite large. For simplicity, you can plot projections of these surfaces in the x-y plane simply by plotting (x,y) as you iterate u and v in a

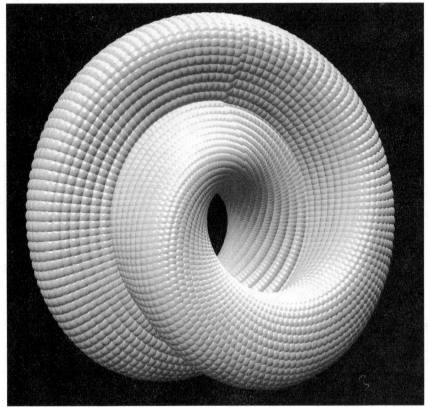

Figure C.1   Banchoff Klein bottle. (Computer rendition by author.)

computer program. Alternatively, here's a handy formula for viewing the curves at any angle:

$$x_p = x \cos \theta + y \sin \theta$$
$$y_p = - x \sin \theta \, \sin \phi + y \cos \theta \, \sin \phi + z \cos \phi$$

where (x, y, z) are the coordinates of the point on the curve prior to projection and (θ, ϕ) are the viewing angles in spherical coordinates.

The Banchoff Klein bottle[1] (Fig. C.1 and C.2) is based on the Möbius band, a surface with only one edge. The Möbius band is an example of a nonorientable space, which means that it is not possible to distinguish an object on the surface from its reflected image in a mirror. This Klein bottle contains Möbius bands and can be built in 4-D space. Powerful graphics computers allow us to design unusual objects such as these and then investigate them by projecting them in a 2-D image. Some physicists and astronomers have postulated that the large-scale structure of

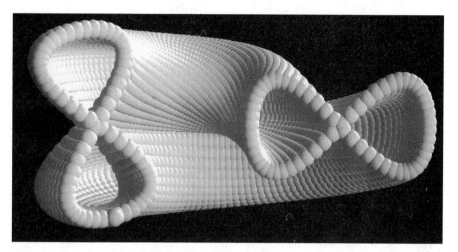

Figure C.2 Cross section of Banchoff Klein bottle, revealing "internal" surfaces. (Computer rendition by author.)

our universe may resemble a huge nonorientable space with Klein bottle-like properties, permitting right-handed objects to be transformed into left-handed ones.

If you are a teacher, have your students design and program their own patterns by modifying the parameters in these equations. Make a large mural of all the student designs labeled with the relevant generating formulas.

ALGORITHM: How to Create a Banchoff Klein Bottle

```
for(u= 0; u < 6.28; u = u + .2) {
  for(v= 0; v < 6.28; v = v + .05 {
    x = cos(u)*(sqrt(2)+cos(u/2)*cos(v)+sin(u/2)*sin(v)*cos(v));
    y = sin(u)*(sqrt(2)+cos(u/2)*cos(v)+sin(u/2)*sin(v)*cos(v));
    z = -sin(u/2)*cos(v)+cos(u/2)*sin(v)*cos(v);
    DrawSphereCenteredAt(x,y,x)
  }
}
```

(The program code here is in the style of the C language.)

quarternions

The invention of quaternions must be regarded as a most remarkable feat of human ingenuity.

—Oliver Heaviside

It is as unfair to call a vector a quaternion as to call a man a quadruped.

—Oliver Heaviside

Some of you may be familiar with the concept of "complex numbers" that have a real and imaginary part of the form $a + bi$, where $i = \sqrt{-1}$. (If you've never heard of complex numbers, feel free to skip this section and simply enjoy the pretty fractal image.) When these 2-D numbers were invented, many people were not sure of their validity. What real-world significance could such imaginary numbers have? However, it didn't take long for scientists to discover many applications for these numbers—from hydrodynamics to electricity.

Quaternions are similar to complex numbers but of the form $a + bi + cj + dk$ with one real and three imaginary parts.[1] The addition of these 4-D numbers is fairly easy, but the multiplication is more complicated. How could such numbers have practical application? It turns out that quaternions can be used to describe the orbits of pairs of pendulums and to specify rotations in computer graphics.

Quaternions are an extension of the complex plane and were discovered in 1843 by William Hamilton while attempting to define 3-D multiplications. Hamilton was a brilliant Irish mathematician whose genius for languages was evident at an early age. He could read at three—by four he had started on Greek, Latin, and Hebrew—and by ten had become familiar with Sanskrit. By age seventeen, his mathematical prowess became evident.

In 1843, during a flash of inspiration while walking with his wife, Hamilton realized that it took four (not three) numbers to accomplish a 3-D transformation of one vector into another. In that instant, Hamilton saw that one number was needed to adjust the length, another to specify the amount of rotation, and two more to specify the plane in which rotation takes place. This physical insight led Hamilton to study hypercomplex numbers (or quaternions) with four components, sometimes written with the form: $Q = a_0 + a_1i + a_2j + a_3k$ where the as are ordinary real numbers, and i, j, and k are each an imaginary unit vector pointing in

Figure D.1   A 2-D slice of a 4-D quaternion Julia set. (Computer rendering by the author.)

three mutually perpendicular directions of space, in a simple extension of ordinary complex numbers of 2-D space. Although it was difficult to visualize quaternions, Hamilton found a way to use them in electrical circuit theory. Oliver Heaviside, a great Victorian-age genius, remarked: "It is impossible to think in quaternions—you can only pretend to do it."

Today, quaternions are everywhere in science. They are used to describe the dynamics of motion in three-space. The space shuttle's flight software uses quaternions in its computations for guidance, navigation, and flight control for reasons of compactness, speed, and avoidance of singularities. Quaternions are used by protein chemists for spatially manipulating models of protein structure. Ted Kaczynski, the Unabomber, spoke of quaternions fondly throughout his highly theoretical mathematical journal articles. Quaternion representations are so complicated that it is useful to develop methodologies to aid in their display. Such methods reveal an exotic visual universe of forms. In particular, I enjoy image processing of the beautiful and intricate structures resulting from the iteration (repeated application) of quaternion equations. Figure D.1 is actually a 2-D slice of a resultant 4-D object called a quaternion Julia set. This slice has a fractional dimension. For details on these 4-D fractal shapes, see note 1 to Appendix D.

appendix e

four-dimensional mazes

So long as we have not become aware that the presence of God is a space, encompassing the whole of reality just as the three-dimensional space does, so long as we conceive the world of God only as the upper story of the cosmic space, so long will God's activity, too, always be a force which affects earthly events only from above.

—Karl Heim, *Christian Faith and Natural Science*

Mazes are difficult to solve in two and three dimensions, but can you imagine how difficult it would be to solve a 4-D maze? Chris Okasaki, from Carnegie Mellon University's School of Computer Science, is one of the world's leading experts on 4-D mazes. When I asked him to describe his 4-D mazes, he replied:

My 4-D mazes are two-dimensional grids of two-dimensional grids. Each of the subgrids looks like a set of rooms with some of the walls missing, allowing the maze-solver to travel directly between certain rooms. In addition, each room may have a set of arrows in it, pointing North, South, East, and West. The arrows mean that you can travel directly between this room and the corresponding room in the next subgrid in that direction. For example, in a 2 × 2 × 2 × 2 maze, if you are in the upper left corner of the upper left subgrid, an arrow pointing south means that you can travel to the upper left corner of the lower left subgrid.

Mathematically, the mazes I generate are based on "random spanning trees" of some graph representing all the possible connections between rooms. Contrary to what you might expect, however, random spanning trees do not make very good mazes. The problem is that they have far too many obvious dead-ends, which do not lure the person solving the maze into exploring them. Therefore, I post-process each random spanning tree to convert a tree with many short dead-ends into one with fewer, longer dead-ends.

A 4-D grid is no harder to model as a graph than a 2-D grid, so my software can generate 4-D mazes just by starting with the appropriate graph. The only difference is in how the resulting spanning tree is displayed. I've also thought about how to do this for 6-D (or even higher) mazes. My visual representation of a 6-D maze is a 2-D grid of 2-D grids of 2-D grids. I use arrows just as in the 4-D maze, except that now arrows can be short or long. A short arrow indicates connections within a 4-D subgrid. A long arrow indicates connections between adjacent 4-D subgrids.

You can find a gallery of mazes, including a random 4-D maze, at http://www.cs.columbia.edu/~cdo/maze/maze.html. Note, however, that this particular 4-D maze does not include the post-processing I mentioned earlier.

Those of you who wish to learn more about 4-D geometry will be interested in *HyperSpace*, a fascinating journal on all subjects relating to higher-dimensional geometries, complex mazes, geometry, and art, and unusual patterns. The journal has articles in English and Japanese. Contact: Japan Institute of Hyperspace Science, c/o K. Miyazaki, Graduate School of Human and Environmental Studies, Kyoto University, Sakyo-ky, Kyoto 606 Japan.

appendix f

smorgasbord for computer junkies

The intersection of a 4-D object with a 3-space does not need to be connected, just like a continuous coral formation can appear as multiple disjoint islands where they protrude above the ocean's surface. A collection of creatures such as a hive of bees may be different parts of a single 4-D animal. Similarly, all people may be part of a single 4-D entity. The aging process can be represented as the slow motion of an intersecting hyperplane through a 4-D entity.

—Daniel Green, Superliminal Software

The presence of God is not an upper story of the one cosmic space, but a separate, all-embracing space by itself, so that the polar and the suprapolar worlds do not stand with respect to one another in the same relation as two floors of the same house but in the relation of two spaces.

—Karl Heim, *Christian Faith and Natural Science*

Code 1. Hyper-hypercube Program

The following is C computer code I wrote to compute the attractive models of higher-dimensional cubes for Figures 4.12 to 4.17. Cubes of dimension N may be generalized to higher dimensions $N + 1$ by translating the N-cube and interconnecting the appropriate vertices—just as a graphical representation of a cube can be generated by drawing two squares and interconnecting the vertices. At higher dimensions, the cubes become so complex that they may be difficult to graphically represent. In the program, $n = 4$ generates a hypercube; $n = 5$ generates a hyperhypercube, and so on. The idea for this approach comes from a BASIC program written by Jonathan Bowen based on a Fortran program by C. S. Kuta. [See, for example, "Hypercubes" in *Practical Computing*, 5(4): 97–99, April, 1982.] For more details, see "On the Trail of the Tesseract," a section in Chapter 4.

```
/* C Program Used to Draw Cubes in Higher Dimensions */
#include <math.h>
#include <stdio.h>
main()
{
float xstart, ystart, x1[10], y1[10], iflag1[10], iflag2[10];
int   i, j, k, n;
float x, y, f, p, c;
        /* n is the dimension of the cube */
        n = 4; p=3.14159/(float)n; i= -1;
        for (j=1; j<=n; j=j+2) {
          i=i+1; c=(float)i*p; x1[j]=cos(c); y1[j]=sin(c);
        }
        i-n;
        for (j=2; j<=n; j=j+2) {
          i=i-1; c=(float)i*p; x1[j]=cos(c); y1[j]=sin(c);
        }
        f=0.0;
        for(j=1; j<=n; j++){f = f + y1[j];}
        xstart=0;
        for(j=1; j<=n; j++){
          if(x1[j]<0) xstart=xstart+x1[j]; iflag1[j]=0;
        }
        ystart=0;
        for(i=1; i<pow(2,n); i++) {
        /* To draw trace of cube before movement,
          use this for loop: for(i=1; i<=pow(2,n-1); i++) { */
          for(j=1; j<=n; j++) {iflag2[j] = iflag1[j];}
          for(j=1; j<=n; j++) {
          if (iflag1[j]==1) goto breaker; iflag2[j]=1; x=0; y=0;
          for(k=1; k<=n; k++) {
            x = x+ iflag1[k]*x1[k]; y = y+ iflag1[k]*y1[k];
            }
          printf("%f %f \n",x*f,y*f); /* first point of line segment */
          x=0; y=0;
          for(k=1; k<=n; k++) {
            x = x + iflag2[k]*x1[k]; y = y + iflag2[k]*y1[k];
            }
          printf("%f %f \n",x*f,y*f); /* second point of line segment*/
```

```
            iflag2[j]=0;
breaker: printf(" ");
} /*j*/
j=1;
breaker2: if(iflag1[j]==0) iflag1[j]=1;
    else {iflag1[j]=0; j++; goto breaker2}
        } /*i*/
}
```

Code 2. Compute the Volume of a Ten-Dimensional Ball

The following code may be used to compute the volume of a ball in any dimensions. The variable "k" is the dimension, in this example, 10. The program is in the style of the REXX language. See Chapter 4 for more information.

```
/* */
pi = 3.1415926; r = 2; k = 10
/* If even dimension: */
IF ((k // 2) = 0) then do
     ans = ((pi ** (k/2.)) * r**k)/factorial(k/2)
END

/* If odd dimension: */
IF ((k // 2) = 1) then do
     m = (k+1)/2; fm = factorial(m); fk = factorial(k+1)
     ans = (pi**((k-1)/2.)* fm * (2**(k+1))*r**k ) / fk
END
say k ans

/* A recursive procedure to compute factorial */
factorial: Procedure
        Arg n
        If n=0 Then Return 1
    Return factorial(n-1)*n
```

Code 3. Draw a Klein Bottle (Mathematica)

Mathematica is a technical software program from Wolfram Research (Champaign, Illinois). With this versatile tool it is possible to draw beautiful Klein bottles as discussed in Chapter 5. The following is a standard recipe for creating Klein bottle shapes using Mathematica.

```
In[1]:= botx = 6 Cos[u] (1 + Sin[u]);
   boty = 16 Sin[u];
   rad = 4 (1 - Cos[u] / 2);

In[4]:= X = If[Pi < u <= 2 Pi, botx + rad Cos[v + Pi],
                     botx + rad Cos[u] Cos[v]];
   Y = If[Pi < u <= 2 Pi, boty, boty + rad Sin[u] Cos[v]];
   Z = rad Sin[v];

In[7]:= ParametricPlot3D[{X, Y, Z}, {u, 0, 2 Pi}, {v, 0, 2 Pi},
                     PlotPoints -> {48,12}, Axes -> False,
                     Boxed -> False, ViewPoint-> {1.5, -2.7, -1.6}]
Out[7]= -Graphics3D-
```

The following is a fragment of code from a C program that draws little spheres along the surface of a Klein bottle.

```
ALGORITHM: How to Create a Klein Bottle
  pi = 3.1415;
  for(u= 0; u<= 2*pi; u = u +.040){
    for(v= 0; v< 2*pi; v= v +.040){
      botx = 6.*cos(u)*(1. + sin(u));
      boty = 16.*sin(u);
      rad = 4.*(1. - cos(u)/2.);
      if ((u > pi)&&(u <= 2*pi)) x = botx + rad*cos(v + pi);
       else x = botx + rad*cos(u)*cos(v);
      if ((u > pi)&&(u <= 2*pi)) y = boty;
       else y = boty + rad*sin(u)*cos(v);
      z = rad*sin(v);
      DrawSphereCenteredAt(x,y,x)
  }
 }
```

appendix g

evolution of four-dimensional beings

The brain acts as a filter of reality, reducing our perception to what is necessary for survival. Therefore, we have developed sensory organs that only perceive 3-D. As our neural pathways form in the first years of life, our brains are programmed to see reality in accord with these organs, and hence we limit it to 3-D. Given all that cannot be understood in three-dimensional terms (Einstein-Podolsky-Rosen Paradox, or even just the simple wave-particle duality), it seems that we are functioning in a reality composed of more than three dimensions. God may be a being in the infinite dimension, or perhaps a being that can conceive of all dimensions easily.

—Lindy Oliver, personal communication

The presence of God, the side of this Power which is turned towards us—and indeed with our human thinking we can never penetrate into the essential nature of this Power—is in fact a *space*. A space, of course, is not a self-contained whole, with definable boundaries separating it in the objective sense from something else.

—Karl Heim, *Christian Faith and Natural Science*

Surface Area and Volume

It is likely that 4-D beings would possess internal organs with some vague similarities to their Earthly counterparts because 4-D creatures will have to perform functions that are carried out most efficiently by specialized tissues. For example, 4-D beings may have digestive and excretory systems, a transport system to distribute nutrients through the body, and specialized organs to facilitate movement. Evolutionary pressure would probably lead to familiar ecological classes of carnivores, herbivores, parasites, and beneficial symbiotic relationships. Technological 4-D beings will have appendages comparable to hands and feet for manipulating objects. Technological beings must also have senses, such as sight, touch, or hearing, although the precise nature of senses that evolve on another world would depend on the environment. For example, some 4-D aliens may have eyes sensitive to the infrared or ultraviolet regions of the spectrum because this has survival value on a particular world. Creatures fulfilling some of these basic trends would be quite different from us, with various possible symmetries, and they could be as big as Tyrannosarus or as small as a mouse depending on gravity and other factors.

We might expect intelligent 4-D beings to have digestive systems resembling a tube structure since we see this so commonly in our world in many different environments. For example, most Earthly animals above the level of cnidarians and flatworms have a complete digestive tract—that is, a tube with *two* openings: a mouth and an anus. There are obvious advantages of such a system compared to a gastrovascular cavity, the pouch-like structure with one opening found in flatworms. For example, with two openings, the food can move in one direction through the tubular system that can be divided into a series of distinct sections, each specialized for a different function. A section may be specialized for mechanical breakdown of large pieces of food, temporary storage, enzymatic digestion, absorption of the products of digestion, reabsorption of water, and storage of wastes. The tube is efficient and has great potential for evolutionary modifications in different environments and for different foods.

As just alluded to, many of our earliest multicelled sea creatures were essentially tubes that could pump water. As life evolved, this basic topological theme did not change—the major structural differences involved complex organs attached to the tube. You and I are still just tubes extending through a body bag filled with "sea water."

In any given dimension, the larger the volume of an animal, the smaller, in proportion, its exposed surface area. As a consequence, larger organisms (in which the interior cells are metabolically active) must increase the surfaces over which diffusion of oxygen and carbon dioxide can occur. These creatures probably must develop a means for transporting oxygen and carbon dioxide to and from this surface area.

In higher dimensions, the fraction of volume near the surface of a bag-like form can increase dramatically compared to 3-D forms. This also implies that most of the blood vessels uniformly distributed through the bag will be close to the surface too (assuming the creatures have blood). As a result, heat dissipation and movement of nutrients and oxygen may be strongly affected by the proximity of the volume to surface. How would this affect the evolution of life? Would primitive 4-D beings evolve so that they digest on the outer surface, while other organs, like brains and hearts (whose primary purpose is not nutrient and oxygen acquisition) are placed deep inside? The outer surface may be digestive and pulmonary, containing sense organs and orifices for excretion and sex. Is this tendency more likely as the dimension of the space increases?[1]

Here is why the volume of a 4-D animal is more concentrated near the surface than in 3-D animals. Consider a *D*-dimensional sphere. The volume is

$$V(r) = S(D) \times r^D/D$$

where S is the so-called "solid angle" in dimension D, and r the radius. (Generally speaking, the solid angle is the set of rays emanating from a point and passing through a particular continuous surface. $S(D)$ is the largest possible measure of a solid angle in D dimensions. The measure of a solid angle is the surface that it intersects of the unit sphere whose center is its vertex.) If we compare two spheres of radius 1 and $1-a$, where a is very small, the difference is

$$(V(1) - V(1-a))/V(1) = 1-(1-a)^D$$

This is essentially the fraction of the volume within $1-a$ of the surface. For example, if you have a 10-D sphere, then about 40 percent of the volume is within $0.05 \times r$ of the surface ($a = 0.05$). In a 4-D sphere, about 34 percent of the volume is within $0.1 \times r$ of the surface. In a 3-D sphere, this is 27 percent. For the $D = 10$ sphere, the fraction within $0.1 \times r$ is 65 percent. (Note that these numbers are much higher for nonspherical shapes.)

Also note that for a given volume V, the surface-area-to-volume ratio increases when going from 3-D to 4-D creatures. This, in turn, implies a favorable oxygen-exchange ratio for respiration and nutrient exchange; it also implies that large animals may be stronger in the fourth dimension, partly because of increased muscular attachment sites. This also means that higher-dimensional beings might be bigger than their 3-D counterparts. In addition, warm-blooded 4-D creatures may need to have efficient means of temperature regulation if the ambient environment has a greater effect on their bodies. If 4-D creatures had different sizes and metabolic rates than us—with accompanying different lifespans and sleep durations—this could make it difficult to communicate with them. (These difficulties might be overcome with "asynchronous" communication such as e-mail.)

Notice that the surface-area-to-volume ratio for a given spatial *extent* increases as one goes from the third dimension to the fourth. We see this easily in the second and third dimensions by using familiar formulas for area and volume for circles and spheres:

Area	Volume	Area/Volume
$2 \times \pi \times r$	$\pi \times r^2$	$2/r$ (circle)
$4 \times \pi \times r^2$	$(4/3) \times \pi \times r^3$	$3/r$ (sphere)

challenging questions for further thought

The unbelievably small and the unbelievably vast eventually meet—like the closing of a gigantic circle. I looked up, as if somehow I would grasp the heavens. The universe, worlds beyond number, God's silver tapestry spread across the night. And in that moment, I knew the answer to the riddle of the infinite. I had thought in terms of man's own limited dimension. I had presumed upon nature. That existence begins and ends is man's conception, not nature's. And I felt my body dwindling, melting, becoming nothing. My fears melted away. And in their place came acceptance. All this vast majesty of creation, it had to mean something. And then I meant something, too. Yes, smaller than the smallest, I meant something, too. To God, there is no zero. I still exist!

—Scott Carey, *The Incredible Shrinking Man*

This appendix includes various unsolved problems, challenges, breakthroughs, and recent news items relating to higher spatial dimensions.

Photons

Is it difficult to imagine a universe with four dimensions in which we are constrained to three dimensions? Could there be some mechanism constraining us— such as being stuck to a surface via adsorption? If we were stuck to some surface in this manner, then the way in which 4-D beings could interact with us would be constrained by that surface.

What happens to photons in four dimensions? For example, how do 4-D beings see? Our 3-D universe doesn't seem to be "losing" photons into the fourth dimension, so how would 4-D creatures use photons to see us or themselves?[1]

Four-Dimensional Speech, Writing, and Art

Would we be able to hear the speech of a 4-D being inches away in the fourth dimension? How would sound work? What would a 4-D being sound like using its 4-D vocal cords?

Perhaps a 4-D being's speech will propagate via 4-D sound (pressure) waves and thus its speech will be partially audible to us. Philosopher Greg Weiss suggests that we would hear something, but not likely the whole message, which may have modulations in 4-D space. A 4-D creature could speak "down" to us, but unless we have some way to manipulate four dimensions, we'd have trouble speaking to it in its language.

Would a being's voice get louder as the creature grew closer to our 3-D space? Would our houses shield our voices from them? If a 4-D being talked to us, would the sound appear to come from everywhere at once, even from inside us?

Of course, we need not use speech to communicate. If we could transmit just a single stream of bits (e.g., ones and zeros) into the fourth dimension, communication would be possible. In fact, the rapid appearance and disappearance of a 4-D object in our world could be used to communicate a binary message.

What would the writing of 4-D beings resemble? Our writing is one dimension lower than the space in which we reside. Would the writing of 4-D beings be three-dimensional? What would this writing look like if projected on a piece of paper in our world (Figure H.1)?

Figure H.2 and H.3 are an artist's renditions of what a 4-D handwriting might look like as it intersected our world. Wild!

As a final artistic piece, consider Figure H.4 by Professor Carlo H. Sequin from the University of California, Berkeley. His representation is a projection of a 4-D 120-cell regular polytope (a 4-D analog of a polygon). This structure consists of twelve copies of the regular dodecahedron—one of the five Platonic solids that exist in 3-D space. This 4-D polytope also has 720 faces, 1200 edges, and 600 vertices, which are shared by two, three, and four adjacent dodecahedra, respectively.

Figure H.1 What would the intersection of Omegamorph handwriting with a piece of paper look like? Would we recognized it as writing?

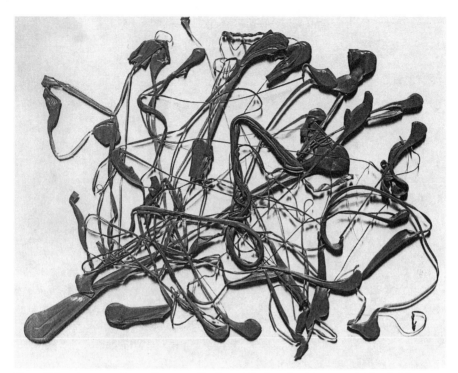

Figure H.2 An artist's renditions of what 4-D handwriting might look like as it intersected our world.

Figure H.3 More fanciful renditions of what 4-D handwriting might look like as it intersected our world.

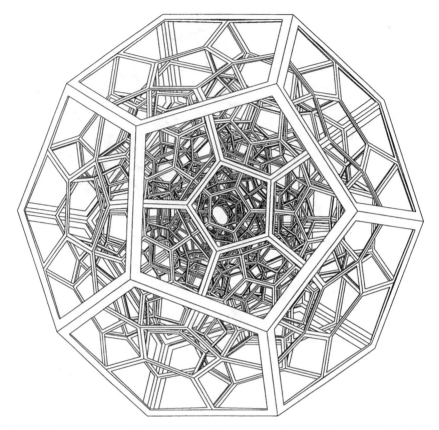

Figure H.4 Projection of a 4-D 120-cell regular polytope. (Courtesy of Professor Carlo H. Sequin.)

The Ultimate Challenge of Living in Other Dimensions

Physicists cannot give a reason why space has three dimensions. Perhaps the dimensionality of space in our universe was "accidentally" determined during the Big Bang, billions of years ago. It does seem that life would be more challenging in other dimensions. As we discussed, it would be difficult for digestive tracts to run through a creature in two dimensions because the tract would cut the creature into two pieces. Richard Morris in *Cosmic Questions* suggests that if the dimensionality of space were four or greater, then stable planetary orbits would not be possible. Morris implies that if a planet did manage to form, it would follow a path that caused it to spiral into the sun. This line of thinking is extended in Max Tegmark's wonderful recent article "On the Dimensionality of Spacetime" appearing in the journal *Classical and Quantum Gravity* (see Further Readings).

Consider a universe with m spatial dimensions and n time dimensions. These universes are classified as $(n + m)$ universes. For example, our universe could be a $(3 + 1)$ universe with three spatial dimensions and one dimension of time. Max Tegmark of the Institute for Advanced Study in Princeton, New Jersey, suggests that all universes—except for a $(3 + 1)$ dimensional universe—may be "dead universes" in the sense that they are devoid of observers. He believes that higher-dimensional spaces cannot contain traditional atoms or perhaps any stable structures. In a space with less than three dimensions, there may be no gravitational force. In universes with more or less than one time dimension, living creatures could not make predictions. These ideas are so fascinating that I would like to explain them just a bit further.

Some kinds of universes are more likely to contain observers than others. Here is some background. As far back as 1917, Paul Ehrenfest suggested that neither classical atoms nor planetary orbits can be stable in a space with $n > 3$. In the 1960s, F. Tangherlini further suggested that traditional quantum atoms cannot be stable in higher dimensional universes (see Further Readings). For physicist readers, these properties are related to the fact that the fundamental Green's functions of the Poisson equation $\nabla^2\phi = \rho$—which gives the electrostatic/gravitational potential of a point particle—is r^{2-n} for $n > 2$. As Tegmark points out, this means that the inverse-square law of electrostatics and gravity become an inverse-cube law if $n = 4$, and so on. When $n > 3$, the two-body problem no longer has any stable orbits as solutions (see I. Freeman's 1969 paper).

In simple English, this implies that if you were in a 4-D universe and launched planets toward a sun, the planets would either fly away to infinity or spiral into the sun. (This is in contrast to a $(3 + 1)$ universe that obviously can, for example, have stable orbits of moons around planets.) A similar problem occurs in quantum mechanics, in which a study of the Schrödinger equation shows that the hydrogen atom has no bound states for $n > 3$. This seems to suggest that it is difficult for higher universes to be stable over time and contain creatures that can make observations about the universe.

Lower dimensional worlds (such as 1-D and 2-D worlds) may not be able to have gravitational forces, as discussed in *Gravitation* by John Wheeler and colleagues and in a paper by Stanley Deser.

So far we have been talking about spatial dimensions, but we may also postulate the existence of different time dimensions. Tegmark believes that a universe will only be able to have observers if there is just one time dimension (i.e., $m = 1$). What would it be like to live in a universe with more than one time-like dimension? Would we have difficulty going through our daily routines of life, job, along with the search for an ideal mate? Even with two or more time dimensions, you might *perceive* time as being one-dimensional, thereby having a pattern of thoughts in a linear succession that characterizes perception of reality. You may travel along an essentially 1-D (time-like) world line through the $(m + n)$ universe. Your wristwatch would work. However, the world would be odd. If two people moving in different time directions happen to meet on the street, they would inevitably drift apart in separate time directions again, unable to stay together! Also, as discussed by J. Dorling, particles like protons, electrons, and photons are unstable and may decay if there is more than one dimension of time.

All sorts of causal paradoxes can arise with more than one dimension of time. However, I do not think this precludes life, even if the behavior or the universe would be quite disturbing to us. Also, electrons, protons, and photons could still be stable if their energies were sufficiently low—creatures could still exit in *cold* regions of universes with greater than one time dimension. However, without well-defined cause and effect in these universes, it might be difficult for brains (or even computers) to evolve and function.

None of these arguments rules out the possibility of life in the fourth spatial dimension (i.e., a $(4 + 1)$ universe). For example, stable structures may be possible if they are based on short distance quantum corrections to the $1/r^2$ potential or on string-like rather than point-like particles.

A Simple Proof That the World Is Three-Dimensional

In 1985, Tom Morley (School of Mathematics, Georgia Institute of Technology) published a paper titled "A Simple Proof That the World Is Three-Dimensional." The paper begins:

> The title is, of course, a fraud. We prove nothing of the sort. Instead we show that radially symmetric wave propagation is possible only in dimensions one and three.

In short, this means that it may be difficult to have radio, television, and rapid global communication in higher-dimensional worlds. [For a theoretical discussion, see Morley, T. (1985) A simple proof that the world is three-dimensional. *SIAM Review*. 27(1): 69–71.]

When I asked Tom Morley about some of the implications of his theories, he replied:

> These results should affect the inhabitants of other dimensions. Surely, Abbott's *Flatland* creatures would have greater challenges than we do when communicating. In three dimensions, sounds get softer as we walk away, but in two dimensions, they get increasingly spread out in

time and space. For example, clap your hands in 2-D, and people far away cannot tell exactly when the clapping started. In a two-dimensional world, the sharp, sudden impulse of the hand clap becomes a "rolling hill" (in a plot of loudness vs. time) that is not well localized in time. It is difficult to tell the precise instant the hand clap occurred. Also, there is not much attenuation of the sound. One can hear the sounds for greater distances in 2-D space than in 3-D space. The sounds also last longer in 2-D space than in 3-D space so that a 1 second sound in 3-D might last for 10 seconds in 2-D. In 4-D there is sufficient attenuation of signals; however, as in the 2-D example, sounds (and all other signals) get "mushed out." Let me give an example. The inhabitants of the fourth dimension could not fully appreciate Beethoven's music because the impressive start to Beethoven's *Fifth Symphony* (dum dum dum DAH) becomes ladidadiladidadidadidididi. Additionally, a four-dimensional creature would find it hard to get a clean start in a 100-meter run by listening to the cracking sound of the starter's gun.

Seven-Dimensional Ice

Recently, scientists and mathematicians have researched the theoretical melting properties of ice in higher dimensions. In particular, mathematicians Nassif Ghoussoub and Changfeng Gui, from the University of British Columbia, have developed mathematical models for how ice turns from solid into liquid in the seventh dimension and have proven that, if such ice exists, it likely exhibits a different melting behavior from ice in lower dimensions. This dependence on dimension, although not very intuitive, often arises in the field of partial differential equations and minimal surfaces—recent results suggest that geometry depends on the underlying dimension in ways that were not suspected in the past. Other research suggests that there is something about 8-D spaces that makes physical phase transitions inherently different from 7-D spaces. If you want to read more about what happens when you lick a 7-D popsicle, see Ekeland, I. How to melt if you must. *Nature.* April 16, 1998, 392(6677): 654–655.

Caged Fleas in Hyperspace

My favorite puzzles involve "flea cages" or "insect cages" for reasons you will soon understand. Consider a lattice of four squares that form one large square:

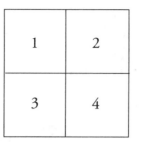

1	2
3	4

How many rectangles and squares are in this picture? Think about this for a minute. There are the four small squares marked "1," "2," "3," and "4," plus two horizontal rectangles containing "1" and "2" and "3" and "4," plus 2 vertical rectangles, plus the one large surrounding border square. Altogether, therefore, there are nine four-sided overlapping areas. The lattice number for a 2 × 2 lattice is therefore 9, or $L(2) = 9$. What is $L(3)$, $L(4)$, $L(5)$, and $L(n)$? It turns out that these lattice numbers grow very quickly, but you might be surprised to realize just how quickly. The formula describing this growth is fairly simple for an $n \times n$ lattice: $L(n) = n^2 (n + 1)^2/4$. The sequence goes 1, 9, 36, 100, 225, 441. . . . For a long time, I've liked to think of the squares and rectangles (quadrilaterals) as little containers or cages in order to make interesting analogies about how the sequence grows. For example, if each quadrilateral were considered a cage that contained a tiny flea, how big a lattice would be needed to cage one representative for each different variety of flea (Siphonaptera) on earth? To solve this, consider that Siphonapterologists recognize 1830 varieties of fleas. Using the equation I have just given you, you can calculate that a small 9 × 9 lattice could contain 2025 different varieties, easily large enough to include all varieties of fleas. (For Siphonaptera lovers, the largest known flea was found in the nest of a mountain beaver in Washington in 1913. Its scientific name is *Hystirchopsylla schefferi*, and it measures up to 0.31 inches in length—about the diameter of a pencil).

It is possible to compute the number of cage assemblies for *3-D* cage assemblies as well. The formula is: $L(n) = ((n^3)(n + 1)^3)/8$. The first few cage numbers for this sequence are: 1, 27, 216, 1000, 3375. Tim Greer of Endicott, New York, has generalized the formula to hyperspace cages of any dimension, m, as $L(n) = ((n^m)(n + 1)^m/ (2^m)$. Let's spend some time examining 3-D cages before moving on to the cages in higher dimensions.

How large a 3-D cage assembly would you need to contain all the species of insects on earth today? (To solve this, consider that there may be as many as thirty million species of insect, which is more than all other phyla and classes put together). Think of this as a zoo where one member of each insect species is placed in each 3-D quadrilateral. It turns out that all you need is a 25 × 25 × 25 ($n = 25$) lattice to create this insect zoo for thirty million species.

To contain the approximately five billion people on earth today, you would need a $59 \times 59 \times 59$ cage zoo. You would only need a $40 \times 40 \times 40$ ($n = 40$) zoo to contain the 460 million humans on earth in the year 1500.

Here is a table listing the size of the cages needed to contain various large numbers, assuming that each quadrilateral contains a single unit of whatever is listed (e.g., pills, objects, stars, or colors):

1. Largest number of objects found in a person's stomach:
   2,533 ($5 \times 5 \times 5$ cage)
   (This number comes from a case involving an insane female who at the age of forty-two swallowed 2,533 objects, including 947 bent pins.)
2. Number of different colors distinguishable by the human eye:
   10 million ($21 \times 21 \times 21$ cage)
3. Number of stars in the Milky Way galaxy:
   10^{12} ($141 \times 141 \times 141$ cage)

Let's conclude by examining the cage assemblies for fleas in higher dimensions. I've already given you the formula for doing this, and it stretches the mind to consider just how many caged fleas a hypercage could contain, with one flea resident in each hypercube or hypertangle.

The following are the sizes of hypercages needed to house the 1,830 flea varieties I mentioned earlier in different dimensions:

Dimension (m)	Size of Lattice (n)
2	9
3	5
4	4
5	3
6	3
7	2

This means that a small $n = 2$, 7-D lattice ($2 \times 2 \times 2 \times 2 \times 2 \times 2 \times 2$) can hold the 1,830 varieties of fleas! An $n = 9$ hyperlattice in the fiftieth dimension can hold each electron, proton, and neutron in the universe (each particle in its own cage).

Here are a few wild challenges.

1. If each cage region were to contain a single snow crystal, what size lattice would you need to hold the number of snow crystals necessary to form the ice age, which has been estimated to be 10^{30} crystals? If you were to draw

this lattice, how big a piece of paper would you need? Provide answers to this question for 2-D and 3-D dimensional figures.

2. If each cage region contained a single grain of sand, what size lattice would you need to hold the number of sand grains contained on the Coney Island beach, which has been estimated to be 10^{20} grains? If you were to draw this lattice, how big a piece of paper would you need? Also provide answers to this question for a hyperlattice in the fourth dimension.

3. Akhlesh Lakhtakia has noted that the lattice numbers $L(n)$ can be computed from triangular numbers $(T_n)^m$. Why should the number of cage assemblies be related to triangular numbers? (The numbers 1, 3, 6, 10, . . . are called triangular numbers because they are the number of dots employed in making successive triangular arrays of dots. The process is started with one dot; successive rows of dots are placed beneath the first dot. Each row has one more dot than the preceding one.)

Optical Aid for "Seeing" Higher Universes

Have you ever wondered what it would really be like to gaze at the fleshy, hairy blobs that this book suggests as models for 4-D beings intersecting our world? Luckily, it is quite easy for students, teachers, and science-fiction fans to gaze at such odd apparitions. Visionary engineer William Beaty gives exact construction details for an optical device that, when pointed at a person's skin (or other body parts), gives a realistic impression of a 4-D being's cross section—namely, as he puts it, "fleshy, pulsating balls of skin, covered with sweaty hair!" At his Internet web page (http://www.eskimo.com/~billb/amateur/dscope.html), Beaty describes the visual effect in detail:

> While looking through the device, I moved my arm up and down. The ball of flesh pulsed. I put the palm of my hand on the end of the mirror device, and made a nice clean smooth sphere of skin. I cupped my hand to fold the skin, and this produced an obscenely throbbing wrinkled flesh-ball. I shoved some fingers into the end, and saw a spiny sphere of waving fleshy pseudopods. I placed it against the side of my fist, clenched and unclenched it, and created throbbing organic orifices. I grabbed coworkers, placed my mouth against the end, made biting and tongue movements, and said, "Look into this thing." They recoiled in revulsion and/or hilarity.

To create the "4-D viewer," he uses three trapezoidal-shaped mirrors, each with dimensions of 12" × 5" × 2" (In other words, the two small edges are 12 inches

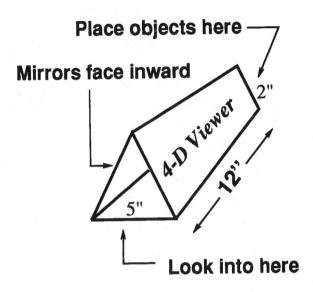

Figure H.5 Optical aid for "seeing" higher universes. (Invention by William Beaty.)

apart as shown in Fig. H.5. You can try other geometries as well, such as a 12" × 7" × 2" arrangement.) The mirrors are taped together at their edges to form a triangular tube with a reflective inner surface. You look into the larger triangular end as you gaze at your hand to see the flesh blobs. (If you were to look into the smaller end at your friend, Beaty says you would see a "spherical glob-monster covered with eyes.") This is a wonderful, fun classroom project, and William Beaty gives practical tips for illumination and safety.

Magic Tesseract

Mathematician John Robert Hendricks has constructed a 4-D tesseract with magic properties. Just as with traditional magic squares whose rows, columns, and diagonals sum to the same number, this 4-D analog has the same kinds of properties in four-space. Figure H.6 represents the projection of the 4-D cube onto the 2-D plane of the paper. Each cubical "face" of the tesseract has six 2-D faces consisting of 3 × 3 magic squares. (The cubes are warped in this projection in the same way that the faces of a cube are warped when draw on a 2-D paper.) To understand the magic tesseract, look at the "1" in the upper left corner. The top forward-most edge contains 1, 80, and 42, which sums to 123. The vertical columns, such as 1, 54, and 68 sum to 123. Each oblique line of three numbers—such as 1, 72, and

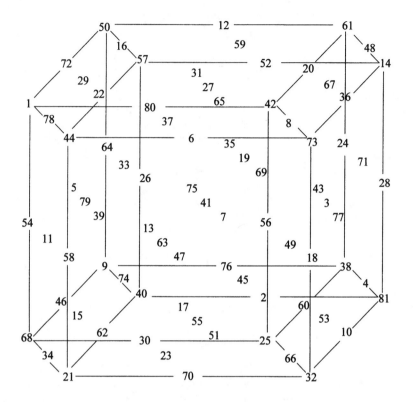

Figure H.6   Magic tesseract by John Robert Hendricks. (Rendering by Carl Speare.)

50—sums 123. A fourth linear direction shown by 1, 78, and 44 sums 123. Can you find other magical sums? This figure was first sketched in 1949. The pattern was eventually published in Canada in 1962 and later in United States. Creation of the figure dispelled the notion that such a pattern could not be made.

HyperDNA

Susana Zanello, Ph.D., from the Boston University School of Medicine, has long pondered the evolution of molecular and cellular processes in hypothetical 4-D beings. She speculates in a letter to me:

> In three-dimensional creatures, the genetic code for the phenotype (the visible properties of an organism, like skin color) exists as a

"string" of information in the DNA. This information can be considered one dimensional, like a sequence of letters in a sentence. After transcription into RNA and translation into proteins, a final folding of the protein structure in three dimensions is necessary for the protein to function properly. With four-dimensional living cells, a similar process includes transcription and translation into a protein folded in the fourth dimension.

In order to correctly form the four-dimensional protein, the original information may be conveyed by a molecule one dimension higher than the one-dimensional DNA sequence. Perhaps this hyperDNA would exist as a 2-D "DNA sheet" which could be schematically represented as an array or matrix of letters. A closed circular surface would offer more combinations or possibilities for DNA information, but this would require more accurate and sophisticated start and stop codons to specify the boundaries of the DNA used for coding a particular protein.

In conceiving a two-dimensional information storage molecule, I search for a shapes that can maximize information storage on a minimal surface as well as make the processes of maintenance and transfer of information (replication) more efficient. For pedagogical reasons, let us imagine a hyperDNA molecule with the same number of information molecules. In our world, DNA contains the basic hereditary information of all living cells and consists of a four-letter code: G, C, A, and T. The letters represent chemical "bases." In a higher universe, we would have a much larger set of possible codons (clusters of bases that code for an amino acid) resulting from the higher-dimensional DNA and the various arrangements of adjacent bases in two dimensions. The coding system would be much more rich.

Highly evolved four-dimensional creatures would have extraordinarily developed nervous systems due in part to the increased number of possible neuronal synapses. Therefore the beings would be super-intelligent. I would expect new and "higher" senses to be present. For example, "propioception," or pressure sensing, would probably not occur perpendicular to a 2-D creature's plane of existence. Similarly, four-dimensional creatures would have senses and sensory receptors that would make little sense to us.

hyperspace titles

Suppose I somehow gain access to a 2-D world, and I convince the 2-D people I am their one true deity. From their point of view, there is only one God. But from my point of view, I am only one of many people who could appear as a God to these people. If God is a multidimensional being, is he alone in his dimension? Could he be only one of many Gods in his realm? If God moved parts of his body in and out of our dimensional realm, could he appear as a pillar of fire or a burning bush?

—Darren Levanian, personal communication

That Christ may dwell in your hearts by faith; that ye, being rooted and grounded in love, May be able to comprehend with all saints what is the breadth, and length, and depth, and height; And to know the love of Christ, which passeth knowledge, that ye might be filled with all the fullness of God.

Ephesians 3:17–19

While writing this book, I performed a computer search of the scientific literature mostly for articles with "hyperspace" in their titles, but also for some with "hyperspace" in the subject matter. A few titles were suggested by colleagues. Consider the following list a random walk through hyperspace to show a variety of applications in physics, mathematics, and computer science.

1. Carter, J. and Saito, M. (1998) *Knotted Surfaces and their Diagrams (Mathematical Surveys and Monographs, No 5)*. New York: American Mathematical Society. (Extends the methods of 3-D topology to 4-D topology.)

2. Devchand, C. and Nuyts, J. (1997) Supersymmetric Lorentz-covariant hyperspaces and self-duality: equations in dimensions greater than four. *Nuclear Physics B.* 503(3): 627–56. (The authors generalize the notions of supersymmetry and superspace by allowing generators and coordinates transforming according to more general Lorentz representations than the spinorial and vectorial ones of standard lore.)

3. Bordley, R. F. (1997) Discrete-time general relativity and hyperspace. *Nuovo Cimento B.* 112B(4): 561–74. (The author describes two popular, but dis-

tinctly different approaches to a unified field theory, including general relativity using ten or more dimensions and discrete-time Lagrangians.)

4. Ryabov, V. A. (1996) Molecular dynamics in curved hyperspace. *Physics Letters A.* 220(4–5): 258–62. (The author describes a new molecular dynamics method considering the crystal as arranged on a curved hypersurface.)

5. Vilenkin, A. (1995) Predictions from quantum cosmology. *Physical Review Letters.* 74: 846–49. (The worldview suggested by quantum cosmology is that inflating universes with all possible values of the fundamental constants are spontaneously created out of nothing. The author explores the consequences of the assumption that we are a "typical" civilization living in this metauniverse.)

6. Morsi, N. N. (1994) Hyperspace fuzzy binary relations *Fuzzy Sets and Systems.* 67(2): 221–37.[The author associates with each implication operator in (0, 1)-valued logic, under certain conditions, an algorithm for extending a fuzzy or ordinary binary relation psi from X to Y, to a fuzzy binary relation from I(X) to I(Y), said to be a fuzzy hyperspace extension of psi.]

7. Pesic, P. D. (1993) Euclidean hyperspace and its physical significance. *Nuovo Cimento B.* 108B, ser. 2(10): 1145–53. (Contemporary approaches to quantum field theory and gravitation often use a 4-D space-time manifold of Euclidean signature called "hyperspace" as a continuation of the Lorentzian metric. To investigate what physical sense this might have, the authors review the history of Euclidean techniques in classical mechanics and quantum theory.)

8. Coxeter, H. (1991) *Regular Complex Polytopes.* New York: Cambridge University Press. (Discusses the properties of polytopes, the 4-D analogs of polyhedra.)

9. Gauthier, C. and Gravel, P. (1991) Discontinuous gauge and particles in multiconnected hyperspace-time. *Nuovo Cimento A.* 104A, ser. 2(3): 325–36. (The authors consider a unified field theory on an 11-D hyperspace-time with a multiconnected extra-space. This setting allows solutions to both the classical cosmological constant problem and the chirality problem.)

10. Folger, T. (1990) Shuffling into hyperspace. *Discover.* 12(1): 66–64. (Mathematical proof of a perfect card shuffle.)

11. Hendricks, J. R. (1990) The magic tesseracts of Order 3 complete. *Journal of Recreational Mathematics.* 22(10): 15–26. (Discusses 4-D analogs of magic squares.)

12. Shepard, S. and Simoson, A. (1989) Scouts in hyperspace (computer game) *Computers & Graphics.* 13(2): 253–60. (The authors describe a one-person,

checkers-like pegboard game in n-space. The goal of the game is to advance a peg as far as possible from an initial configuration of pegs. Using an argument based on the golden mean, the authors demonstrate bounds for how far a peg can travel as well as how many pegs are needed to achieve a particular goal. Finally, they view the game as automata moving about so as to achieve a collective goal.)

13. Burton, R. P. (1989) Raster algorithms for Cartesian hyperspace graphics. *Journal of Imaging Technology.* 15(2): 89–95. (The authors design algorithms for Cartesian hyperspace graphics. The hidden volume algorithm clips and performs volume removal in four dimensions. The shadow algorithm constructs shadow hypervolumes that it intersects with illuminated hypersurfaces. The shading algorithm performs solid shading on hyperobjects. The raytracing algorithm introduces a viewing device model that projects from 4-D space to 2-D space.)

14. Linde, A. and Zelnikov, M. (1988) Inflationary universe with fluctuating dimension. *Physics. Letters B* (Netherlands) 215: 59–63. (The authors argue that in an eternal chaotic inflationary universe, the number of uncompactified dimensions can change locally. As a result, the universe divides into an exponentially large number of independent inflationary domains [miniuniverses] of different dimension.)

15. Bertaut, E. F. (1988) Euler's indicatrix and crystallographic transitive symmetry operations in the hyperspaces $E(n)$. *Comptes Rendus de l'Academie des Sciences, Serie II (Mecanique, Physique, Chimie, Sciences de l'Univers, Sciences de la Terre).* 307(10): 1141–46. (The author uses elementary number theory in this article on crystallographic symmetry operations.)

16. Finkelstein, D. (1986) Hyperspin and hyperspace. *Physical Review Letters.* 56(15): 1532–33. (A spinorial time-space $G(N)$ that supports a Kaluza-Klein theory of gauge potentials can be made from N-component spinors of $SL(N,C)$ in the same way that the Minkowskian manifold $G(2)$ is made from two-component spinors of $SL(2,C)$. Also discusses photons and gravitons.)

17. Deser, S., Jackiw, R., and 'tHofft, G. (1984) Three-dimensional Einstein gravity: dynamics of flat space. *Annals of Physics.* 152: 220–35. (In three spacetime dimensions, the Einstein equations imply that source-free regions are flat.)

18. Mei-chi, N., Burton, R. P., and Campbell, D. M. (1984) A shadow algorithm for hyperspace: calculating shadows in hyperdimensional scenes. *Computer Graphics World.* 7(7): 51–59. (The authors develop a shadow algorithm for hyperspace while creating computer graphics techniques for

meaningfully presenting hyperdimensional models that occur whenever four or more variables exist simultaneously.)

19. Lowen, R. (1983) Hyperspaces of fuzzy sets. *Fuzzy Sets and Systems.* 9(3): 287–311.

20. Condurache, D. (1981) Symbolic representation of signals on hyperspaces. II. The response of linear systems to excitations representable symbolically. *Buletinul Institutului Politehnic din Iasi, Sectia III (Electrotehnica, Electronica, Automatizari).* 27(3–4): 49–56. (Describes conditions in which multiplication, raising to a power, and inversion modify the partition class of the elements entering those operations. The results are useful in studying the response of linear systems to symbolically representable excitations.)

21. Cerin, Z. T. and Sostak, A. P. (1981) Fundamental and approximative uniformity on the hyperspace. *Glasnik Matematicki, Serija III.* 16(2): 339–59. [Introduces the fundamental uniformity $2(f, V)$ and the approximate uniformity $2(a, V)$ on the hyperspace $2(x)$ of all nonempty compact subsets of a uniform space (X, V) that reflect space properties of elements of $2(x)$.]

22. Condurache, D. (1981) Symbolic representation of signals on hyperspaces. I. Symbolic representation of modulated signals. *Buletinul Institutului Politehnic din Iasi, Sectia III (Electrotehnica, Electronica, Automatizari).* 27(1–2): 33–42. (Discusses the notion of symbolic representation of a real, derivable function of a scalar argument by means of a finite dimension algebra element on the real number field.)

23. Burton, R. P. and Smith, D. R. (1982) A hidden-line algorithm for hyperspace. *SIAM Journal on Computing.* 11(1): 71–80. (The authors design an object-space hidden-line algorithm for higher-dimensional scenes. Scenes consist of convex hulls of any dimension, each compared against the edges of all convex hulls not eliminated by a hyperdimensional clipper, a depth test after sorting, and a minimax test. Hidden and visible elements are determined in accordance with the dimensionality of the selected viewing hyperspace. The algorithm produces shadows of hyperdimensional models, including 4-D space-time models, hyperdimensional catastrophe models, and multivariable statistical models.)

24. Goodykoontz, J. T., Jr. (1981) Hyperspaces of arc-smooth continua. *Houston Journal of Mathematics.* 7(1): 33–41. (Discusss the hyperspace of closed subsets.)

25. Nadler, S. B., Jr., Quinn, J. E., and Stavrakas, N. M. (1977) Hyperspaces of compact convex sets, II. *Bulletin de l'Academie Polonaise des Sciences. Serie des Sciences Mathematiques, Astronomiques et Physiques.* 25(4): 381–85. [The

authors show that cc($R(n)$) is homeomorphic to the Hilbert cube minus a point.]

26. Kuchar, K. (1976) Dynamics of tensor fields in hyperspace, III. *Journal of Mathematical Physics.* 17(5): 801–20. (Discusses hypersurface dynamics of simple tensor fields with derivative gravitational coupling. The spacetime field action is studied and transformed into a hypersurface action. The hypersurface action of a covector field is cast into Hamiltonian form. Generalized Hamiltonian dynamics of spacetime hypertensors are discussed; closing relations for the constraint functions are derived.)

27. Kuchar, K. (1976) Kinematics of tensor fields in hyperspace, II. *Journal of Mathematical Physics.* 17(5): 792–800. (Differential geometry in hyperspace is used to investigate kinematical relationships between hypersurface projections of spacetime tensor fields in a Riemannian spacetime.)

28. Kuchar, K. (1976) Geometry of hyperspace, I. *Journal of Mathematical Physics.* 17(5): 777–91. (The author defines hyperspace as an infinite-dimensional manifold of all space-like hypersurfaces drawn in a given Riemannian spacetime.)

29. Shu-Chung Koo (1975) Recursive properties of transformation groups in hyperspaces. *Mathematical Systems Theory.* 9(1): 75–82. [Let (X, T) be a transformation group with compact Hausdorff phase space X and arbitrary acting group T. There is a unique uniformity Omega of X that is compatible with the topology of X.)

30. Whiston, G. S. (1974) Hyperspace (the cobordism theory of spacetime). *International Journal of Theoretical Physics.* 11(5): 285–88 (A compact space- and time-orientable spacetime is cobordant in the unoriented sense—that is, it bounds a compact five-manifold. The bounding property is a direct consequence of the triviality of the Euler number.)

31. Tashmetov, U. (1974) Connectivity of hyperspaces. *Doklady Akademii Nauk SSSR.* 215(2): 286–88. (Results regarding connected and locally connected compacts in a hyperspace are extended to the case of arbitrary, full metric spaces.)

32. Caywood, C. (1988) The package in hyperspace. *School Library Journal.* 35(3): 110–11.

33. Easton, T. (1988) The architects of hyperspace. *Analog Science Fiction-Science Fact.* 108(5): 182-83.

34. Boiko, C. (1986) Danger in hyperspace (play). *Plays.* 45: 33–40.

notes

Only God truly exists; all other things are an emanation of Him, or are
His "shadow."

—Afkham Darbandi and Dick Davis,
Introduction to *The Conference of the Birds*

Preface

1. Islamic mystics have historically considered our world a cubic cage, a six-sided
prison. Humans struggle in vain to escape the bondage of the senses and physical
world. Persian poets refer to such imprisonment as "six-door," or *shishdara*—the hope-
less position of a gambler playing a form of backgammon. In this game, players use a
six-sided die to move pieces. If an opposing player has locked (occupied) all six locations to which your playing piece could potentially have moved, then you are "shish-
dar," or six-out, since your piece cannot move.

2. "I want to know if humankind's Gods could exist in the fourth dimension."
Although a 4-D being would have God-like powers, this is not to say a 4-D being would
have *all* the properties we traditionally attribute to God. Could a non-omnipotent, non-
omniscient, non-omnipresent 4-D being act in such a way to *appear* to be a God? What
Judeo-Christian miracles couldn't a 4-D being perform? A 4-D being would probably
not be able to do any of the following, which many religions attribute to God: creating
the world or universe, understanding our inner thoughts and prayers, seeing all temporal
events at once, healing complex biochemical diseases, and prophesizing the future. However, a 4-D being could appear to be omnipresent and in several places at one time if it is
a very large creature. It could probably fool ancient people into accepting it as a God.

It would be difficult for a 4-D creature to masquerade as a 3-D creature; as the 4-D
being moved, its intersection shape in our world would change. (Imagine how difficult
it would for us to masquerade as a 2-D creature for Flatlanders living in a 2-D world.)
Philosopher Greg Weiss wrote to me, "Perhaps this is why 'God' is so uneager to show
himself in the Old Testament!"

Greg Weiss suggests that it is amusing to propose that all spiritual beings, angels,
devils, and God are four-dimensional, but only God has higher powers or access to
even higher dimensions. Heaven could be essentially a 4-D locale in which God, the
devil, and angels can move around at will. But, of course, this is mere speculation,
more in the realm of religion than science.

Chapter 1

1. I asked Professor Michio Kaku, author of *Hyperspace*, if the gravitational curvature of space implies the existence of a fourth dimension. He responded:

> We do not need a fourth spatial dimension in which to describe the curvature of space. From one point of view, the fourth spatial dimension is fictitious. This is because we can use "intrinsic" 3-D coordinates in which the only coordinates are three bent spatial dimensions and one time dimension. Thus, an ant on an ordinary balloon can only see two dimensions, and says that the third dimension in unnecessary, because the ant cannot travel in the third dimension, which is fictitious from his point of view.
>
> However, we can also use "extrinsic" coordinates in which to visualize the bending of space, by embedding spacetime within a higher dimension. All the fancy graphical representations—of black holes as holes and funnels in space, inflating balloons representing the Big Bang, and 99.99% of all the pictures found in general relativity books—are done in extrinsic coordinates. We see the 3-D balloon from the vantage point of a fictitious fourth dimension, which has no physical reality. (After all, if the balloon is the entire universe, then where are you standing when you look at the balloon? You are standing in a fictitious fourth dimension.)
>
> That being said, let me now say that current thinking in theoretical physics postulates the existence of not just four spatial dimensions (one being fictitious), but ten physical dimensions of space and time.
>
> It is confusing that people use the word hyperspace to refer to both: the fictitious fourth spatial dimension used in extrinsic coordinates (essentially a gimmick in which to "see" balloons and holes in space) found in ordinary general relativity, and also higher physical dimensions in which superstrings live.

2. Kaluza-Klein theory (named after two European scientists) suggests the existence of additional dimensions that are rolled up or "compactified" in such a way that they are undetectable at macroscopic levels.

3. On the other hand, Joe Lykken of the Fermi National Accelerator Laboratory in Batavia, Illinois, believes physicists may be able to find experimental evidence for string theory. Although the extra dimensions of strings are compactified—that is, curled up on scales of just 10^{-33} centimeters, which would be out of the reach of any conceivable experiment—Lykken and several other groups are considering the possibility that a few of those dimensions could unravel slightly, opening up onto scales that precision measurements in accelerators or even on a benchtop might actually probe. [For more information on practical tests of string theory, see Kestenbaum, D. (1998) Practical tests for an 'untestable' theory of everything? *Science*. 281(5378): 758–59.]

4. Unfortunately, there are so many different ways to create universes by compactifying the six dimensions that string theory is difficult to relate to the real universe. In 1995, researchers suggested that if string theory takes into account the quantum effects of charged mini black holes, the thousands of 4-D solutions may collapse to only one. Tiny black holes, with no more mass than an elementary particle, and strings may be two descriptions of the same object. Thanks to the theory of mini black holes, physicists now hope to mathematically follow the evolution of the universe and select one particular Calabi-Yau compactification—a first step to a testable "theory of everything."

Some trivia: Gabriele Veneziano, in the late 1960s, worked on string theories. However, interest in his particular version of the theory faded when other physicists showed they would only work in twenty-six dimensions. Also, some researchers believe that all known elementary particles have unseen symmetric twins called *sparticles*.

5. What we've learned in the 20th century is that the great ideas in physics have geometric foundations. —Edward Witten, *Scientific American*

6. Just like the early years of Einstein's theory of relativity, string theory is simply a set of clever equations waiting for experimental verification. Unfortunately, it would take an atom smasher thousands of times as powerful as any on Earth to test the current version of string theory directly. It is hoped that, humans will refine the theory to the point where it can be tested in real-world experiments. With Edward Witten directing his attention to string theory, the world hopes that he and his colleagues can crack the philosophical mystery that's dodged science ever since the ancient Greeks: What is the ultimate nature of the universe? What is the loom on which God weaves?

Whatever that loom is, it has created a structurally rich universe. Most astronomers today believe that the universe is between eight and twenty-five billion years old, and has been expanding outward ever since. The universe seems to have a fractal nature with galaxies hanging together in clusters. These clusters form larger clusters (clusters of clusters). "Superclusters" are clusters of these clusters of clusters. In recent years, there have been other baffling theories and discoveries. Here are just a few:

- In our universe a *Great Wall* exists consisting of a huge concentration of galaxies stretching across 500 million light-years of space.
- In our universe a *Great Attractor* exists, a mysterious mass pulling much of the local universe toward the constellations Hydra and Centaurus.
- There are *Great Voids* in our universe. These are regions of space where few galaxies can be found.
- *Inflation theory* continues to be important in describing the evolution of our universe. Inflation theory suggests that the universe expanded like a drunken balloon-blower's balloon while the universe was in its first second of life.
- The existence of *dark matter* also continues to be hypothesized. Dark matter may consist of subatomic particles that may account for most of the universe's mass. We don't know what dark matter is composed of, but theories include: neutrinos

(subatomic particles), WIMPs (weakly interacting massive particles), MACHOs (massive compact halo objects), black holes, and vast, filamentary networks of warm, ionized gases that are difficult to detect with present satellites.

• *Cosmic strings and cosmic textures* are hypothetical entities that distort the space-time fabric.

(Perhaps some of the large, extragalactic structures are artifacts of inexact observations and analysis. Further research is required to be certain.)

What are the biggest questions for today's scientists? Perhaps, "Are there higher spatial dimensions?" or "Which laws of physics are fundamental and which are accidents of the evolution of this particular universe?" or "Does intelligent, technologically advanced life exist outside our Solar System?" and "What is the nature of consciousness?"

How many of these questions will we ever answer?

7. M-theory, like string theory, relies heavily on the idea of *supersymmetry* in which each known particle having integer spin has a counterpart with the same mass but half-integer spin. Supersymmetry predicts "supergravity" in which a graviton (with spin 2) transmits gravitational interactions and has a partner gravitino with spin 3/2. Conventional gravity does not place constraints on the possible dimensions of spacetime, but with supergravity there is an upper limit of eleven dimensions of spacetime. In 1984, 11-D supergravity theories were disbanded in favor of superstring theory in ten dimensions.

M-theory in eleven dimensions gives rise to the five competing string theories in ten dimensions (two heterotic theories, Type I, Type IIA, and Type IIB). When the extra dimension curls into a circle, M-theory yields the Type IIA superstring. On the other hand, if the extra dimension shrinks to a line segment, M-theory yields one of the heterotic strings.

New theories deal with arcane concepts difficult for mere mortals to grasp. For example, the strength with which objects interact (their charges) is related to the size of invisible dimensions; what is charge in one universe may be size in another; and certain membranes may be interpreted as black holes (or "black-branes") from which nothing, not even light, can escape. The mass of a black-brane can vanish as the hole it wraps around shrinks, allowing one spacetime with a certain number of internal holes (resembling a piece of cheese) to change to another with a different number of holes, violating the laws of classical topology.

Edward Witten and Petr Horava have recently shown how to shrink the extra dimension of M-theory into a segment of a line. The resulting structure has two 10-D universes (each at an end of the line) connected by a spacetime of eleven dimensions. Particles (and strings) exist only in the parallel universes at the ends, which can communicate with each other only via gravity. For additional reading on these mind-numbing concepts, see Michael Duff's paper cited in Further Readings.

Note that string theory says little about the space in which strings move and vibrate. A relatively new mathematical model known as loop quantum gravity represents an alternate approach in which the rules of quantum mechanics are applied directly to Einstein's description of space and time. In this model, space itself comes packaged in tiny discrete units. To quantize space, physicists postulate discrete states analogous to the energy levels or orbitals of atoms. For further reading on quantized space and on other theories including 4-D spin foam, see Ivars Peterson's 1998 *Science News* article.

8. According to a theory devised by James Hartle and Stephen Hawking, time may lose its ordinary, time-like character near the origin of the universe. In their theory, time resembles a spatial dimension at very early "times." Thus the universe has no real beginning for the simple reason that, if one goes sufficiently far back, there are no longer three dimensions of space and one of time, but only four space-like dimensions. In other words, time does not "keep on going," but instead becomes something other than time when one explores the far past. Here, time cooperates with the three spatial dimensions to create a 4-D sphere. At this point, time becomes "imaginary."

Similarly, time may have no end. If the universe eventually contracts back on itself, it may never get to the final singularity because time will become imaginary again.

If the universe has no beginning and no end, we can't ask why it was created at a particular moment in time—because time ceases to exist. (For more details, see Richard Morris's book *Cosmic Questions* in Further Readings.)

Chapter 2

1. However, would a 4-D man be interested in a woman who would seem paper-thin to him?

2. If a 2-D man has a self-gripping gut, how could his brain on one side of the body control the other side? What 2-D chemicals would he use for energy?

3. Actually, they would step *around* the walls by moving a very short distance into the fourth dimension.

4. However, you could still introduce bacteria carried on your 4-D tools. Other advantages would be the minimal damage, healing time, blood loss, and scarring.

5. What motive would a 4-D Don Juan have? Insubstantial women with incompatible genes and ova would not seem like desirable targets.

6. Also consider that in Pointland, there is nothing else in the universe that an inhabitant could imagine needing or wanting.

7. How large would 2-D brains have to be to contain the same number of synapses as in our brains? How could the nerves interconnect without interference or using a lot more area for wiring?

Chapter 3

1. About a year after Morris and Thorne delved into wormholes, Matt Visser of Washington University developed a wormhole model that "looked" more like a rectangular spool of thread than the hourglass shape of Morris and Thorne. A rectangular

hole in the middle of the spool corresponded to the wormhole, the gateway between two regions of space. In this model, described in *Physical Review*, the wormhole's boundaries are straight and can be made as distant from one another as desired. Exotic matter could therefore be placed far away from the wormhole travelers to minimize risks. This rectangular spool model may be more stable than the Morris-Thorne model, and gravitational tidal forces on passengers would be less of a concern. How would the end of a Visser wormhole appear to you as it floated in space? It would look like a dark, rectangular box. You could approach this black prism and enter it near the center. Almost immediately, you would exit a similar dark prison constituting the other side of the wormhole. These two prisms would be connected via the forth dimension along the shaft of the spool that compases the wormhole's throat.

The exterior of the Visser wormhole acts like a giant mirror. Light shining on it would bounce off as if it hit a reflecting material. Visser also proposed another mathematical model for a wormhole that resembles two coreless apples. The inner walls of the fruit are connected along the fourth dimension. You can read more about this structure in Halpern's *Cosmic Wormhole* book or in Visser's original scientific paper.

The Morris-Thorne wormholes can be used for space travel and time travel as described in the paper by Morris, Thorne, and Yurtsever (see Further Readings). Various debates continue about the theoretical possibility of the wormhole time machine. For readers interested in other discussions, see Visser's papers. In my book *Black Holes: A Traveler's Guide*, I also include color graphics of 3-D versions of quantum foam and wormholes.

2. You might try using plexiglass sheets, suspended at corners and center with knotted strings.

3. Go is a more difficult game than chess, but extension to higher dimensions is, in some ways, conceptually simple.

Chapter 4

1. *Self-reproducing universes:* Physicist Andrei Linde's theory of self-reproducing universes implies that new universes are being created all the time through a budding process. In this theory, tiny balls of spacetime called "baby universes" are created in universes like our own and evolve into universes resembling ours. This theory does not mean we can find these other universe by traveling in a rocket ship. These universes that bud off from our own might pinch off from our spacetime and then disappear. (For a very brief moment, a thin strand of spacetime called a wormhole might connect the baby and parent universes. The wormholes would have diameters 10^{20} smaller than the dimension of an atomic nucleus, and the wormhole might remain in existence for only 10^{-43} seconds.) The baby universes would also have offspring, and all the count-

less universes could be very different. Some might collapse into nothingness quickly after their creation. Stephen Hawking has suggested that subatomic particles are constantly traveling through wormholes from one universe to another.

The universe is its own mother: Physicists Li-Xin Li and J. Richard Gott III of Princeton University suggest the possibility of "closed timelike curves"—where there is nothing in the laws of physics that prevents the universe from creating itself! In a 1998 *Science News* article, Gott suggests, "The universe wasn't made out of nothing. It arose out of something, and that something was itself. To do that, the trick is time travel." Li and Gott suggest that a universe undergoing the rapid early expansion known as inflation could give rise to baby universes, one of which (by means of a closed time-like curve) would turn out to be the original universe. "The laws of physics may allow the universe to be its own mother."

The multiverse: In 1998, Max Tegmark, a physicist at the Institute for Advanced Study at Princeton, New Jersey, used a mathematical argument to bolster his own theory of the existence of multiple universes that "dance to the tune of entirely different sets of equations of physics." The idea that there is a vast "ensemble" of universes (a multiverse) is not new—the idea occurs in the many-worlds interpretation of quantum mechanics and the branch of inflation theory suggesting that our universe is just a tiny bubble in a tremendously bigger universe. In Marcus Chown's "Anything Goes," in the June 1998 issue of *New Scientist*, Tegmark suggests that there is actually greater simplicity (e.g., less information) in the notion of a multiverse than in an individual universe. To illustrate this argument, Tegmark gives the example of the numbers between 0 and 1. A useful definition of something's complexity is the length of a computer program needed to generate it. Consider how difficult it could be to generate an arbitrarily chosen number between 0 and 1 specified by an infinite number of digits. Expressing the number would require an infinitely long computer program. On the other hand, if you were told to write a program that produced *all* numbers between 0 and 1, the instructions would be easy. Start at 0, step through 0.1, 0.2, 0.3, and so on, then 0.01, 0.11, 0.21, 0.31. . . . This program would be simple to write, which means that creating all possibilities is much easier than creating one very specific one. Tegmar extrapolates this idea to suggest that the existence of infinitely many universes is simpler, less wasteful, and more likely than just a single universe.

2. Hinton coined the word "tesseract" for the unfolded hypercube in Figure 4.7. Others have used it to mean the central projection in Figure 4.5, while still others use it interchangeably with the word "hypercube." One of the earlier published hypercube drawings (as in Fig. 4.5) was drawn by architect Claude Bragdon in 1913 who incorporated the design in his architecture.

3. Could you really see all thirty-two vertices at once or would you see up to sixteen at a time as the 5-D cube rotated?

4. Now that your mind has been stretched to its limit, I give some interesting graphical exercises.

- A good student exercise is to draw a graph of $y = a^n/n!$ for a fixed a. You'll see the same kind of increase in y followed by a decrease as you do for hyperspheres.
- Draw a 3-D plot showing the relationship between sphere hypervolume, dimension, and radius.
- Plot the *ratio* of a k-dimensional hypersphere's volume to the k-dimensional cube's volume that encloses the hypersphere. Plot this as a function of k. (Note that a box with an two-inch-long edge will contain a ball of radius one inch. Therefore, for this case, the box's hypervolume is simply 2^k.) Here's a hint: It turns out that an n-dimensional ball fits better in an n-dimensional cube than an n-cube fits in an n-ball, if and only if n is eight or less. In nine-space (or higher) the volume ratio of an n-ball to an n-cube is smaller than the ratio of an n-cube to an n-ball.
- Plot the ratio of the volumes of the $(k + 1)$-th dimensional sphere to the kth-dimensional sphere for a given radius r.
- For more technical readers, compute the hypervolume of a fractal hypersphere of dimension 4.5. To compute factorials for non-integers, you'll have to use a mathematical function called the "gamma function." The even and odd formulas given in this chapter yield the same results by interpreting $k! = \Gamma(k + 1)$.
- Can you derive a formula for the surface area of a k-dimensional sphere? How does surface area change as you increase the dimension?

Chapter 5

1. Astronomers actively search for evidence of the universe's shape by looking at detailed maps of temperature fluctuations throughout space. These studies are aided by the sun-orbiting Microwave Anisotropy Probe spacecraft and the European Space Agency's Planck satellite. In a closed, "hyperbolic" universe, what astronomers might think is a distant galaxy could actually be our own Milky Way—seen at a much younger age because the light has taken billions of years to travel around the universe. Cambridge University's Neil Cornish and other astronomers suggest that "if we are fortunate enough to live in a compact hyperbolic universe, we can look out and see our own beginnings."

According to Einstein's theory of general relativity, the overall density of our universe determines both its fate and its geometry. If our universe has sufficient mass, gravity would eventually collapse the universe back in a Big Crunch. In effect, such a universe would curve back on itself to form a closed space of finite volume. The space is said to have "positive curvature" and resembles the surface of a sphere. A rocket traveling in a straight line would return to its point of origin. If our universe had less mass, the universe would expand forever while its rate of expansion gets closer and closer to zero. The geometry of this universe is "flat" or "Euclidean." If the universe had even less mass, the universe expands forever at a constant rate. This kind of space is called "hyperbolic." It has a negative curvature and a shape resembling the seat of a saddle. Currently, observational data suggests that the universe does not have enough mass to make it closed or flat (although recent evidence that neutrinos possess slight mass could affect this because their gravitational effects may mold the shape of galaxies or even possibly reverse the expansion of the universe.) Many astronomers hope for a flat cosmos because it is closely tied to "inflation theory"—a popular conjecture that the universe underwent an early period of rapid expansion that amplified random sub-atomic fluctuations to form the current structures in our universe. In much the same way that expansion makes a small region of a balloon look flat, inflation would stretch the universe, smoothing out any curvature it might have had initially. It is astonishing that we live in an age that all these conjectures will soon be testable with satellites scanning the universe's microwave background radiation. For example, in a hyperbolic universe, strong temperature variations in the microwave background should occur across smaller patches of the heavens than in a flat universe. (See Ron Cowen's and Ivars Peterson's 1998 *Science News* articles in Further Readings.)

Recent computer simulations suggest the existence of vast filamentary networks of ionized gas, or plasma—a cosmic cobweb that now links galaxies and galaxy clusters. These warm cobwebs may be difficult to detect with current satellites. [For more information, see Glanz, J. (1998) Cosmic web captures lost matter. *Science*. June 26, 280(5372): 2049–50.]

Trillions of neutrinos are flying through your body as you read this. Created by the Big Bang, stars, and the collision of cosmic rays with the earth's atmosphere, neutrinos outnumber electrons and protons by 600 million to 1. [For more information, see. Gibbs, W. (1998) A massive discovery. *Scientific American*. August, 279(2): 18–19.]

It is possible that the universe has a strange topology so that different parts interconnect like pretzel strands. If this is the case, the universe merely gives the illusion of immensity and the multiple pathways allow matter from different parts of the cosmos to mix. The July 1998 *Scientific American* speculates that, in the pretzel universe, light from a given object has several different ways to reach us, so we should see several copies of the object. In theory, we could look out into the heavens and see the Earth. [For further reading, see Musser, G. (1998) Inflation is dead: long live inflation. *Scientific American*. 279(1): 19–20.]

2. For some beautiful computer-graphics renditions and explanations of these kinds of surfaces, see Thomas Banchoff's *Beyond the Third Dimension*. For Alan Bennett's spectacular glass models of Klein bottles and variants, see Ian Stewart's March 1998 article in *Scientific American* 278(3): 100–101. Bennett, a glassblower from Bedford, England, reveals unusual cross sections by cutting the bottles with a diamond saw. He also creates amazing Klein bottles with three necks, sets of bottles nested inside one another, spiral bottles, and bottles called "Ouslam vessels" with necks that loop around

twice, forming three self-intersections. (The Ousalm vessel is named after the mythical bird that goes around in ever decreasing circles until it disappears up its own end.) If the Ouslam vessel is cut vertically, it falls apart into two three-twist Möbius bands. Normally, a Klein bottle falls apart into two one-twist Möbius bands when cut, but Bennet and Stewart show that you can also cut a Klein bottle along a different curve to get just one Möbius band. Glassblower Alan Bennett can be contacted at Hi-Q Glass, 2 Mill Lane, Greenfield, Bedford, UK MK45 5DF.

Concluding Remarks

1. See note 1 in Chapter 5.

Appendix A

1. *Solution to the random walk question.* Mathematical theoreticians tell us that the answer is one—infinite likelihood of return for a 1-D random walk. If the ant were placed at the origin of a two-space universe (a plane), and then executed an infinite random walk by taking a random step north, south, east, or west, the probability that the random walk will eventually take the ant back to the origin is also one—infinite likelihood. Our 3-D world is special: 3-D space is the first Euclidean space in which it is possible for the ant to get hopelessly lost. The ant, executing an infinite random walk in a three-space universe, will eventually come back to the origin with a 0.34 or 34 percent probability. In higher dimensions, the chances of returning are even slimmer, about $1/(2n)$ for large dimensions n. The $1/(2n)$ probability is the same as the probability that the ant would return to its starting point on its second step. If the ant does not make it home in early attempts, it is probably lost in space forever. Some of you may enjoy writing computer programs that simulate ant walks in confined hypervolumes and making comparisons of the probability of return. By "confined" I mean that the "walls" of the space are reflecting so that when the ant hits them, the ant is, for example, reflected back. Other kinds of confinement are possible. You can read more about higher dimensional walks in Asimov, D. (1995) There's no space like home. *The Sciences.* September/October 35(5): 20–25.

2. *Solution to the Rubik's tesseract question.* The total number of positions of Rubik's tesseract is 1.76×10^{120}, far greater than a billion! The total number of positions of Rubik's cube is 4.33×10^{19}. If either the cube or the tesseract changed positions every second since the beginning of the universe, they would still be turning today and not have exhibited every possible configuration. The mathematics of Rubik's tesseract are discussed in Velleman, D. (1992) Rubik's tesseract, *Mathematics Magazine.* February 65(1): 27–36.

Appendix C

1. For more information on the Banchoff Klein bottle, see Thomas Banchoff's exquisite book *Beyond the Third Dimension*. More accurately, the bottle is actually constructed from a figure-eight cylinder that passes through itself along a segment. To produce the Banchoff Klein bottle, you twist the figure-eight cylinders as you bring the two ends outward.

Appendix D

1. Quaternions define a 4-D space that contains the complex plane. As we stated previously, quaternions can be represented in four dimensions by $Q = a_0 + a_1 i + a_2 j + a_3 k$ where i, j and k are (like the imaginary number i) unit vectors in three orthogonal directions; they are also perpendicular to the real axis. Note that to add or multiply two quaternions, we treat them as polynomials in i, j, and k, but use the following rules to deal with products: $i^2 = j^2 = k^2 = -1$, $ij = -ji = k$, $jk = -kj = i$, $ki = -ik = -ik = j$.

To produce the pattern in this section, "mathematical feedback loops" were used. Here one simply iterates $F(Q,q)$: $Q \rightarrow Q^2 + q$ where Q is a 4-D quaternion and q is a quaternion constant. The following computer algorithm for squaring a quaternion involves keeping track of the four components in the formula.

```
ALGORITHM — Compute quaternion, main computation
```

```
Variables:
a0,a1,a2,a3,rmu are the real, i, j, and k coefficients
Notes:
This is an 'inner loop' used in the same spirit as in traditional
Julia set computations. No complex numbers are required for the
computation. Hold three of the coefficients constant and examine
the plane determined by the remaining two. This code runs in a
manner similar to other fractal-generating codes in which color
indicates divergence rate. rmu is a quaternion constant.
```

```
DO i = 1 to NumberOfIterations
    savea0 =          a0*a0 - a1*a1 - a2*a2 - a3*a3 + rmu0;
    savea1 =          a0*a1 + a1*a0 + a2*a3 - a3*a2 + rmu1;
    savea2 =          a0*a2 - a1*a3 + a2*a0 + a3*a1 + rmu2;
    savea3 =          a0*a3 + a1*a2 - a2*a1 + a3*a0 + rmu3;
    a0=savea0;a1=savea1;a2=savea2;a3=savea3;
    if (a1**2+a2**2+a3**2+a0**2) > CutoffSquared then leave loop;
end; /*i*/
PlotDot(Color(i));
```

Shades of gray indicate the mapping's rate of explosion. As is standard with Julia sets, "divergence" is checked by testing whether Q goes beyond a certain threshold, τ, after many iterations. For Figure D.1, the mapping is iterated 100 times and the iteration count, n, is stored when $|Q| \sim \tau$. The *logarithm* of the iteration counter, n, is then mapped to intensity in the picture. (See my book *Computers, Pattern, Chaos, and Beauty* for more information.) Figure D.1 represents a 2-D slice of a 4-D quaternion Julia set. The slice is in the (a_0, a_2) plane at level $(a_1, a_3) = (0.05, 0.05)$. The constant $q = (-0.745, 0.113, 0.01, 0.01)$ and $\tau = 2$. The initial value of Q is $(a_0, a_2, 0.05, 0.05)$, where a_0 and a_2 correspond to the pixel position in a figure.

Appendix G

1. In the fourth dimension, is it possible that many new elements would exist and we would be adding an entire dimension to the periodic table? Philosopher Ben Brown believes that there would be many more elements. If we assume that carbon is the only material capable of building life, in four dimensions there would be a greatly decreased chance that the correct elements would come together to form biomolecules compared to 3-D space—thereby slowing down evolutionary processes. The initial life-forms would have virtually no competition, but would take much longer to come into being. The odds of *intelligent* life evolving in a 4-D universe would be even slimmer. The lower the dimension, the better the chance of evolving life because the chances are greater for having the right elements present at the right time.

Four-dimensional beings may well take in oxygen and nutrients from their outer surfaces. Ben Brown believes that a 4-D creature would have no lungs, and would be insect-like. As background, consider that insects cannot grow very large because they have no lungs. Certain insect species breathe through the body wall, by diffusion, but, in general, insects' respiratory systems consist of a network of tubes, or tracheae, that carry air throughout the body to smaller tubelets or tracheoles that all the organs of the body are supplied with. In the tracheoles, the oxygen from the air diffuses into the bloodstream, and carbon dioxide from the blood diffuses into the air. The exterior openings of the tracheae are called spiracles. In four dimensions, a being's surface area would be so much greater in proportion to its internal volume that body size would be less limited. Perhaps many creatures would be "flat" to take full advantage of the extended surface area.

Most of a 4-D creature's energy could be taken directly through a 4-D network of pores, rather than ingested, because little space would exist inside for organs like a brain and heart. The creature's nervous system might operate using electrical signals oscillating through three dimensions and traveling through a fourth, creating a highly complex synaptic web. The heart would have to work very hard if it were centrally located and had to pump blood over great distances. Perhaps 4-D creatures would have no central hearts but rather a mass of arteries to push blood along.

Ben Brown suggests that 4-D beings would be solar powered. In three dimensions one photon contains 10^7 ergs; this energy would be significantly higher in four dimensions, due to a possible added oscillation. This means that one photon could be a more significant source of energy. In our 3-D world, it seems impractical for a mobile, warm-blooded creature to use solar energy; however, in four dimensions, it may be practical for beings to absorb light and turn it directly into a storable form of energy. Any such stored energy might reside as tissues deep inside the creature, shielding vital organs such as the brain.

Appendix H

1. Obviously it is quite difficult to predict how particles will behave in different dimensions. This means that it is difficult to predict what senses would evolve in 4-D beings. In three dimensions, light oscillates in a 2-D fashion while moving forward through the third dimension (the electrical and magnetic fields oscillate at right angles to one another). In four dimensions, would light somehow oscillate in three dimensions while moving forward through a fourth? Of course this is quite speculative, but it may mean that "3-D light" would not be visible to a 4-D being, and that 2-D light may not be visible to us.

If "light" could exist in two dimensions, "light" might oscillate through one dimension and move forward through a second. Does this mean that Flatland might not be visible to us as we stood above Flatland and looked down? Perhaps this implies that 4-D creatures would have difficulty seeing us. What would we see if our eyes were level with Flatland? (Would we be able to see a Flatlander if we illuminated him with a 3-D light source?) If we were able to see Flatlanders, movement and energy stored on the 2-D surface might somehow be translated into the energy levels of 3-D particles.

Arlin Anderson believes that sound (vibrational energy) seems to dissipate all its energy within our 3-D world. This implies that a 4-D being can't hear (or see) us with our sound or light until the being puts his ear or eye on our level. However, if 4-D photons and phonons are possible, the being could use these to observe us.

further readings

Higher space can be viewed as a background of connective tissue tying together the world's diverse phenomena.

—Rudy Rucker, *The Fourth Dimension*

The identification of the omnipresence of space with the omnipresence of God leads to a serious difficulty.

—Max Jammer, *Concepts of Space*

Abbott, E. (1952) *Flatland*. New York: Dover. [The original publication was in 1884 (Seeley & Co.), and the most recent Dover Thrift Edition appeared in 1992.]

Apostol, T. (1969) *Calculus*, Volume II, 2d. Edition. New York: John Wiley & Sons.

Banchoff, T. (1996) *Beyond the Third Dimension: Geometry, Computer Graphics, and Higher Dimensions, 2d Edition*. New York: Freeman.

Banchoff, T. (1990) From Flatland to Hypergraphics. *Interdisciplinary Science Reviews*. 15:364–72.

Berlinghoff, W. and Grant, K. (1992) *Mathematics Sampler: Topics for Liberal Arts*. New York: Ardsley House Publishing.

Bond, N. (1974) "The monster from nowhere," in *As Tomorrow Becomes Today*, Charles W. Sullivan, ed. New York: Prentice-Hall. (Originally published in *Fantastic Adventures*, July 1939.)

Buchel, W. (1963) Why is space three dimensional? (in German) *Physikalische Blätter*. 19:547–48.

Cowen, R. (1998) Cosmologists in Flatland: Searching for the missing energy. *Science News*. February 28, 153(9):139–41.

Deser, S., Jackiw, R., and 'tHofft, G. (1984). Three-dimensional Einstein gravity: dynamics of flat space. *Annals of Phyics*. 152:220–35.

Dewdney, A. (1984) *The Planiverse: Computer Contact with a Two-Dimensional World*. New York: Poseidon.

Dorling, J. (1969) The dimensionality of time. *American Journal of Physics*. 38:539–42.

Duff, M. (1998) The theory formerly known as strings. *Scientific American*. February, 278(2):64–69. (Discusses membrane theory.)

Dyson, F. (1978) Characterizing irregularity. *Science*. May 12, 200(4342):677–78.

Dyson, F. (1979) Time without end: physics and biology in an open universe. *Reviews of Modern Physics*. 51(3):447–60.

Ehrenfest, P (1917) Can atoms or planets exist in higher dimensions? *Proceedings of the Amsterdam Academy.* 20:200–203.

Everett, H. (1957) Relative state formulation of quantum mechanics. *Reviews of Modern Physics.* 29 (July):454–62.

Freeman, I. (1969) Why is space three-dimensional? *American Journal of Physics.* 37:1222–24.

Friedman, N. (1998) Hyperspace, hyperseeing, and hypersculpture. Preprint available from Professor Nat Friedman, Mathematics Department, University of Albany SUNY, Albany, New York 12222.

Gardner, M. (1990) *The New Ambidextrous Universe.* New York: Freeman.

Gardner, M. (1982) *Mathematical Circus.* New York: Penguin.

Gardner, M. (1969) *The Unexpected Hanging.* New York: Simon and Schuster.

Gardner, M. (1965) *Mathematical Carnival.* New York: Vintage.

Halpern, P. (1993) *Cosmic Wormholes.* New York: Plume.

Heim, K. (1953) *Christian Faith and Natural Science.* New York: Harper and Row. (Reprinted in 1971 by Peter Smith, Glouchester, Mass.)

Heinlein, R. (1958) "—And he built a crooked house." In *Fantasia Mathematica*, C. Fadiman, ed. New York: Simon and Schuster. (Story originally published in 1940.)

Hendricks, J. (1990) The magic tesseracts of Order 3 complete. *Journal of Recreational Mathematics.* 22(1):16–26.

Hendricks, J. (1962) The five-and six-dimensional magic hypercubes of Order 3. *Canadian Mathematical Bulletin.* May, 5(2):171–89.

Henricks, J. (1995) Magic tesseract. In *The Pattern Book: Fractals, Art, and Nature*, C. Pickover, ed. River Edge. N.J.: World Scientific.

Horgan, J. (1991) The Pied Piper of superstrings. *Scientific American.* November 265(5):42–44.

Kaku, M. (1994) *Hyperspace.* New York: Oxford University Press.

Kasner, E. and Newman, R. (1989) *Mathematics and the Imagination.* New York: Tempus. (A reprint of the 1940 edition.)

Misner, C. W., Thorne, K. S., and Wheeler, J. A. (1973) *Gravitation.* New York: Freeman. (Excellent comprehensive background for more technical readers. A megatextbook on Einstein's theory of relativity, among other things. Lots of equations.)

Morris, M. S. and Thorne, K. S. (1988) Wormholes in spacetime and their use for interstellar travel: A tool for teaching general relativity. *American Journal of Physics.* 56:395.

Morris, M. S., Thorne, K. S., and Yurtsever, U. (1988) Wormholes, time machines, and the weak energy conditions. *Physical Review Letters.* 61:1446.

Morris, R. (1993) *Cosmic Questions.* New York: John Wiley & Sons.

Pappas, T. (1990) *More Joy of Mathematics.* San Carlos, Calif.: Wide World Publishing/Tetra.

Peterson, I. (1998) Circle in the sky: detecting the shape of the universe. *Science News.* February, 153(8):123–35.

Pickover, C. (1998) *Black Holes: A Traveler's Guide.* New York: John Wiley & Sons.

Peterson, I. (1998) Evading quantum barrier to time travel. *Science News.* April 11, 153(19):231.

Peterson, I. (1998) Loops of gravity: calculating a foamy quantum space-time. *Science News.* June 13, 153(24):376–77.

Rucker, R. (1984) *The Fourth Dimension.* Boston: Houghton-Mifflin.

Rucker, R. (1977) *Geometry, Relativity, and the Fourth Dimension.* New York: Dover.

Stewart, I. (1998) Glass Klein bottles. *Scientific American.* 278(3):100–101.

Tangherlini, F. (1963) Atoms in higher dimensions. *Nuovo Cimento.* 27:636–639.

Tegmark, M. (1997) On the dimensionality of spacetime. *Classical and Quantum Gravity.* 14:L69–L75.

Thorne, K. S. (1994) *Black Holes and Time Warps: Einstein's Outrageous Legacy.* New York: W. W. Norton.

Velleman, D. (1992) Rubik's tesseract. *Mathematics Magazine.* February 65(1): 27–36.

Visser, M. (1989) Traversable wormholes from surgically modified Schwarzchild space-times. *Nuclear Physics.* B328:203.

Visser, M. (1989) Traversable wormholes: some simple examples. *Physical Review.* 39D:3182.

Visser, M. (1990) Wormholes, baby universes, and causality. *Physical Review.* 41D: 1116.

about the author

Clifford A. Pickover received his Ph.D. from Yale University's Department of Molecular Biophysics and Biochemistry. He graduated first in his class from Franklin and Marshall College, after completing the four-year undergraduate program in three years. He is author of the popular books *The Science of Aliens* (Basic Books, 1998), *Strange Brains and Genius* (Plenum, 1998), *Time: A Traveler's Guide* (Oxford University Press, 1998), *The Alien IQ Test* (Basic Books, 1997), *The Loom of God* (Plenum, 1997), *Black Holes: A Traveler's Guide* (John Wiley & Sons, 1996), and *Keys to Infinity* (Wiley, 1995). He is also the author of numerous other highly acclaimed books including *Chaos in Wonderland: Visual Adventures in a Fractal World* (1994), *Mazes for the Mind: Computers and the Unexpected* (1992), *Computers and the Imagination* (1991) and *Computers, Pattern, Chaos, and Beauty* (1990), all published by St. Martin's Press—as well as the author of over 200 articles concerning topics in science, art, and mathematics. He is also coauthor, with Piers Anthony, of *Spider Legs*, a science-fiction novel recently listed as Barnes & Noble's second best-selling science-fiction title.

Pickover is currently an associate editor for the scientific journals *Computers and Graphics, Computers in Physics,* and *Theta Mathematics Journal,* and is an editorial board member for *Odyssey, Speculations in Science and Technology, Idealistic Studies, Leonardo,* and *YLEM.* He has been a guest editor for several scientific journals. He is the editor of *Chaos and Fractals: A Computer-Graphical Journey* (Elsevier, 1998), *The Pattern Book: Fractals, Art, and Nature* (World Scientific, 1995), *Visions of the Future: Art, Technology, and Computing in the Next Century* (St. Martin's Press, 1993), *Future Health* (St. Martin's Press, 1995), *Fractal Horizons* (St. Martin's Press, 1996), and *Visualizing Biological Information* (World Scientific, 1995), and coeditor of the books *Spiral Symmetry* (World Scientific, 1992) and *Frontiers in Scientific Visualization* (John Wiley & Sons, 1994), Dr. Pickover's primary interest is finding new ways to continually expand creativity by melding art, science, mathematics, and other seemingly disparate areas of human endeavor.

The *Los Angeles Times* recently proclaimed, "Pickover has published nearly a book a year in which he stretches the limits of computers, art and thought." Pickover received first prize in the Institute of Physics' "Beauty of Physics Photographic Competition." His computer graphics have been featured on the covers of many popular magazines, and his research has recently received considerable attention by the press—including CNN's "Science and Technology Week," The Discovery

Channel, *Science News*, the *Washington Post*, *Wired*, and the *Christian Science Monitor*—and also in international exhibitions and museums. *OMNI* magazine recently described him as "Van Leeuwenhoek's twentieth-century equivalent." *Scientific American* several times featured his graphic work, calling it "strange and beautiful, stunningly realistic." Pickover has received numerous U.S. patents, including Patent 5,095,302 for a 3-D computer mouse, 5,564,004 for strange computer icons, and 5,682,486 for black-hole transporter interfaces to computers.

Dr. Pickover is currently a Research Staff Member at the IBM T. J. Watson Research Center, where he has received sixteen invention achievement awards, three research division awards, and four external honor awards. Dr. Pickover is also lead columnist for the brain-boggler column in *Discover* magazine.

Dr. Pickover's hobbies include the practice of Ch'ang-Shih Tai-Chi Ch'uan and Shaolin Kung Fu, raising golden and green severums (large Amazonian fish), and piano playing (mostly jazz). He is also a member of the SETI League, a group of signal-processing enthusiasts who systematically search the sky for intelligent, extraterrestrial life. Visit his web site, which has received over 200,000 visits: http://sprott.physics.wisc.edu/pickover/home.htm. He can be reached at P.O. Box 549, Millwood, New York 10546-0549 USA.

addendum

As this book goes to press, I have uncovered several recent discussions dealing with parallel universes and cosmic topology.

- *Parallel Universes*—Chapter 3 discussed parallel universes and the "many-worlds" interpretation of quantum mechanics. Readers interested in a lively and critical discussion of this topic should consult Professor Victor Stenger's *The Unconscious Quantum* (Prometheus Books, 1995). For example, he doubts very much that the parallel universes (in the many-worlds interpretation) all simultaneously exist. He also does not believe that all branches taken by the universe under the act of measurement are "equally real." Stenger discusses other approaches such as the "alternate histories" theory that suggests every allowed history does not occur. What actually happens is selected by chance from a set of allowed probabilities.

- *More on the Multiverse*—Many have wondered why the hypothetical cosmological constant (a mysterious energy that seems to be permeating space and counteracting gravity on cosmic distance scales) is just right to permit life in our universe. Some theories, in fact, predict that the constant should be much larger and therefore would presumably keep galaxies, stars, and life from forming. Uncomfortable with the idea that the cosmological constant and other parameters are simply lucky accidents, Stephen Hawking recently suggested that an infinity of big bangs have gone off in a larger "multiverse," each with different values for these parameters. Only those values that are compatible with life could be observed by beings such as ourselves. For more information, see James Glanz, "Celebrating a century of physics, en masse," *Science*, 284(5411): 34–35, 1999.

- *Cosmic topology*—Note 1 for chapter 5 discusses various topologies for our universe. The April, 1999 issue of *Scientific American* suggests the universe could be spherical yet so large that the observable part seems Euclidean, just as a tiny patch on a balloon's surface looks flat. In other topologies, the universe might be "multiply connected" like a torus, in which case there are many different direct paths for light to travel from a source to an observer.

Many of you are probably asking: what is outside the universe? The answer is unclear. I must reiterate that this question supposes that the ultimate physical reality must be a Euclidean space of some dimension. That is, it presumes that if space is a hypersphere, then the hypersphere must sit in a four-dimensional Euclidean space, allowing us to view it from the outside. As the authors of the *Scientific American* article point out, nature need not adhere to this notion. It would be perfectly acceptable for the universe to be a hypersphere and not be embedded in any higher-dimensional space. We have difficulty visualizing this because we are used to viewing shapes from the outside. But there need not be an "outside."

As many papers have been published on cosmic topology in the past three years as in the preceding 80. Today, cosmologists are poised to determine the topology of our universe through observation. For example, if we look out into space, and images of the same galaxy are seen to repeat on rectangular lattice points, this will suggest we live in a 3-torus. (Your standard 2-torus, or doughnut shape, is built from a curved square, while a 3-torus is built from a cube.) Sadly, finding such patterns would be difficult because the images of a galaxy would depict different points in time. Astronomers would need to be able to recognize the same galaxy at different points in its history.

One of the most difficult ideas to grasp concerning cosmic topology is how a hyperbolic space can be finite. For more information on this and other topics in cosmic topology, see: Jean-Pierre Luminet, Glenn D. Starkman, and Jeffrey R. Weeks, "Is space infinite?" *Scientific American,* April 280(4): 90–97, 1999

index